Uprisings

Uprisings is a wonderful title for a wonderful book. Sarah Simpson and Heather McLeod do a great job of telling the story of the community grain revolution that is spreading across North America. They show how each uprising is helping to build a more sustainable agriculture that is all about community resilience and local food security. Their hands-on instructions for growing your own grains is very informative and well-researched. *Uprisings* provides lots of loaves of inspiration and knowledge to enable you and your community to join the revival of small-scale grain production.

—Dan Jason, Salt Spring Seeds

For those of us who are convinced that food independence is the key to preserving personal freedom as well as political and economic freedom, *Uprisings* is an extremely important book about the cultural and economic situation society faces today. It chronicles the first courageous steps toward food independence in growing grains and processing them locally into good food for humans. It describes in detail both the startup successes and sometimes failures in this effort and in the second part, gives excellent instruction in how to grow and process grains. Together, the two parts deliver a clear writing on the wall of how society can (and must) decentralize food production and thwart the dangerous monopolies that now threaten the food market.

—Gene Logsdon, author, *The Eternal Garden*

In *Uprisings*, McLeod and Simpson reveal not only their passion for locally grown food, and in particular grains, but a compelling case for home gardeners to think of grain crops as they do traditional garden vegetables. This book is filled with meticulously researched case histories that will prove invaluable to any smaller-scale farmer of grains. The authors manage to demystify these most basic of food crops with elegant descriptions and clear instructions. They provide a plan for all of us, from backyard gardeners to small organic growers, to discover for ourselves free and sustainable access to a commodity that industry might normally remove from our reach. In *Uprisings*, a new standard has been set on the topic of growing, harvesting, and processing grains, and the Field-to-Table movement is stronger because of this book.

—Mark Macdonald, West Coast Seeds

We've lost our way with raising grains, and *Uprisings* has come to put us back on the path to good health and honest self-sufficiency. With this book, Simpson and McLeod have gifted the reader with a tool chest of skills for resiliency and self-reliance that runs the gamut from the basics of heirloom grains to the value of community... with a wealth of solid how-to for small-scale local grain production in between.

—Richard Freudenberger, publisher, *BackHome* magazine, and author, *Alcohol Fuel: Making and Using Ethanol as a Renewable Fuel*

Uprisings

A Hands-on Guide to the Community Grain Revolution

Sarah Simpson & Heather McLeod

new society
PUBLISHERS

Cover design by Diane McIntosh.
© istock : Hands (Angie Photos) Wheat (small-frog) Borders – Ulimi

Printed in Canada. First printing September 2013

New Society Publishers acknowledges the support of the Government of Canada through the Book Publishing Industry Development Program (BPIDP) for our publishing activities.

Paperback ISBN: 978-0-86571-734-3   eISBN: 978-1-55092-542-5

Inquiries regarding requests to reprint all or part of *Uprisings* should be addressed to New Society Publishers at the address below.

To order directly from the publishers, please call toll-free (North America) 1-800-567-6772, or order online at www.newsociety.com

Any other inquiries can be directed by mail to:

New Society Publishers
P.O. Box 189, Gabriola Island, BC V0R 1X0, Canada
(250) 247-9737

New Society Publishers' mission is to publish books that contribute in fundamental ways to building an ecologically sustainable and just society, and to do so with the least possible impact on the environment, in a manner that models this vision. We are committed to doing this not just through education, but through action. The interior pages of our bound books are printed on Forest Stewardship Council®-registered acid-free paper that is **100% post-consumer recycled** (100% old growth forest-free), processed chlorine free, and printed with vegetable-based, low-VOC inks, with covers produced using FSC®-registered stock. New Society also works to reduce its carbon footprint, and purchases carbon offsets based on an annual audit to ensure a carbon neutral footprint. For further information, or to browse our full list of books and purchase securely, visit our website at: **www.newsociety.com**

Library and Archives Canada Cataloguing in Publication

Simpson, Sarah, 1978-, author

Uprisings : a hands-on guide to the community grain revolution / Sarah Simpson & Heather McLeod.

Includes bibliographical references and index.

ISBN 978-0-86571-734-3 (pbk.)

1. Flour mills--Canada. 2. Flour mills--United States. 3. Farms, Small--Canada. 4. Farms, Small--United States. 5. Grain--Canada. 6. Grain--United States. 7. Bread. 8. Food security. I. McLeod, Heather, 1980-, author II. Title.

SB189.S54 2013   633.1   C2013-903594-X

Contents

Acknowledgments

OUR THANKS GO FIRST TO THE MANY INDIVIDUALS across North America who generously shared their tales of revolution with us, and who are now quoted in these pages. We hope their stories will inspire a new wave of uprisings.

We share our appreciation and admiration for contrary farmer Gene Logsdon and seed guru Dan Jason, who have been spreading the gospel of small-scale grain production for generations and instigating revolutions in countless communities. Without them, Island Grains would never have been born and we would not have had the opportunity to write this book.

For revolution to succeed, a community must come together. We thank our fellow grainies, who made Island Grains such a wonderful experience and left us wanting more. We thank our husbands, Kevin and Brock, for their encouragement and honest feedback on draft after draft. We're grateful to Maeve Maguire for her brilliant insights on the early chapters, Laura Floyd for coming to the rescue with her adept research skills and Barbara Graves for extolling the virtues of sticktoitiveness.

We will always count our blessings for Heather Nicholas of New Society Publishers. Heather saw a need for this book before we did, and helped us earn the opportunity to write it. We sincerely thank everyone at NSP for their enthusiasm and patience.

Finally, we thank everyone who has ever planted a kernel of grain, milled local wheat into flour, baked or eaten real bread and so furthered the cause.

Preface: A Call to Arms

G RAIN HAS ALWAYS BEEN AN IMPORTANT PART OF THE HUMAN STORY.

Over the past 9,500 years, humankind has progressed from cultivating wild grasses with the largest seeds, to building ovens and baking bread, to inventing water mills, threshing machines, high-tech combines and other specialized equipment. As baker Jonathan Stevens of Hungry Ghost Bread notes, "That whole process — and all the people it takes to cooperate and make that happen and figure out all those bits and pieces — is not only a metaphor for civilization: it *is* civilization."

For most of the many thousands of years that humans have grown grain, each family or community would grow enough to meet its own needs — primarily because the machinery and transportation technology didn't exist to allow us to grow, harvest or ship larger-scale crops.

But that changed in North America in the latter half of the 19th century. Industrial agriculture and cross-continental railways centralized grain production in the Prairie provinces and central states, where the landscape, soil and climate were ideal for efficient grain production.

Because of this cheap, reliable source of grain to supply our kitchens, we didn't just stop growing grain on the rest of the continent . . . many of us forgot that we had ever grown it in the first place. Scythes, sickles, fan mills and other staple equipment for small-scale grain production were left to rust in barns.

Today, North Americans aren't only ignorant of how to grow our own grain — most of us don't even know how to bake a loaf of bread. Growing grains and baking bread were basic skills for much of human history, and still are in many other parts of the world. Losing such skills has made us increasingly dependent on the mainstream food supply.

In the late 1800s, the whole grain flour that was a staple of North American homesteads underwent its own transformation as traditional stone mills began to be replaced by steel rollers. While steel rollers can process more grain in less time, they also produce more heat, which can make the fat in the kernel oxidize and turn rancid. Stone mills simply grind the entire kernel, but steel rollers remove the germ and bran, where much of the grain's fiber and nutrition are stored. This was good news in the 19th century because the remaining white, starchy flour was better able to survive long-distance transportation by railway and had a longer shelf life once it reached grocery stores.

After World War II, Canada and the United States began to enrich this steel-rolled flour with folic acid and other additives. The popularity of flour made it an excellent vehicle for feeding nutritional supplements to citizens.

Meanwhile, the wheat plant itself was bred and transformed to such an extent that, these days, our bodies barely recognize what we call "wheat," and some of us can no longer digest it. The industrial agriculture system has taken over every step of the seed-to-bread process.

That helplessness many of us feel when watching the sad stories of environmental, economic and social disaster on the evening news is a valid warning sign. Losing our connection with "the staff of life" is a symptom of a larger issue. Historically, whoever controlled the grain supply (and the tools for processing it) controlled the people. We've lost our control. No wonder modern-day issues, from climate change to economic recession, leave us feeling overwhelmed: we don't even know what's in our bread these days.

Given the importance of grain to the human story, perhaps it should be no surprise that many of us who want to make a difference in the world — environmentally, economically, socially — are choosing grain as our instrument of change. When a community reclaims control over its food supply, there are numerous ripple effects and spin-off benefits. Eaters fall in love with, and choose to purchase, local food. Farmers gain a new market due to increased demand and enjoy the higher profit margins of selling directly to consumers. The environmental costs of transportation are eliminated, and the local economy improves because food dollars are staying in the community. Ultimately, the community becomes stronger.

The following pages tell the stories of communities from all corners of North America that decided to take back control of their daily bread. Bakers, farmers, eaters, environmentalists and entrepreneurs led these projects. Some wanted to reduce their carbon footprint, while others sought to grow their local economy. Some farmers wanted a profitable crop, while others focused on educating their customers. Some wanted to increase food security, others to connect with their community. For the most part, their strategies were successful.

While their reasons were all unique, some common themes resonate through all these tales from the frontlines of the community grain revolution.

For instance, New Brunswick's Speerville Flour Mill and the Kootenay Grain CSA prove the power we all have as eaters who vote for the food systems we want with every trip to the grocery store, restaurant order and home-cooked meal.

Many of us share a longing for community and connection. The Island Grains education project and Skowhegan, Maine's annual Kneading Conference both tapped into this need when they brought like-minded community members together to share information and learn together.

The Alaska Flour Company and Hungry Ghost Bread's Little Red Hen Project demonstrate how community sufficiency can be even more fulfilling, efficient and effective than self-sufficiency, and that the secret to local food security is to figure out how you can play a part in your community's larger plan.

Through unusual projects such as Arizona's Heritage Grains Collaborative and the Mendocino Grain Project in California's wine country, we see how creative thinking will help us find the loopholes in our current industrial food system.

With uprisings in progress from the snowy fields of Alaska to the southern Arizona desert, we hope these stories inspire you and give you the information you need to see possibilities in your own community, to connect with other like-minded "grainies" and to start your own community grain project.

While reading, you may find yourself craving a fresh loaf of artisan bread or some quality time in your garden. If that's the case, turn to the hands-on guide section of this book, which includes an abundance of information on how to grow grains on a small scale, as well as some of the tastiest, easiest grain-based recipes from our kitchens. You're only a few grain seeds away from starting your own uprising.

Read the book. Grow grain. Join the revolution.

Local grains taste different. They taste like success.
They taste like optimism. They taste like revolution.

— Chris Hergesheimer, the Flour Peddler

Section I

Tales from the Front Lines

Grist for the Mill —
Vote with Your Fork

THE SUCCESS OF ANY BUSINESS DEPENDS ON ITS CUSTOMERS. This is just as true for a farm as it is for a hardware store. It doesn't matter how cooperative the weather is, or how well a farmer plans, or whether they weed at exactly the right time. In the end, if not enough people buy the farm's products, the farm can't survive as a financially sustainable business.

For this reason, farmer entrepreneurs are always trying to figure out what their customers want. Do consumers prefer salad greens or lettuce? Turkey or chicken? With limited time and land, farmers need to grow the right crops, and once they get their wares to the farmers market or grocery store, they need someone to buy them.

In a farmer's world, the customer is everything. Yet, many of us think we are powerless. We're not. We hold all the power.

The 100-Mile Diet, James B. MacKinnon and Alisa Smith's book about their attempt to eat only food produced within 100 miles of their Vancouver apartment, triggered a renaissance of local eating. And yet, while on a book tour in August 2007, MacKinnon told a Creston Valley audience: "We had no idea that we could have that kind of direct impact just as consumers on what's being grown."

Amazing things can happen when consumers realize we have power.

Michael Pollan tells readers to vote with their forks in his book *The Omnivore's Dilemma*. Buying kale at a farm stand, for example, isn't just a healthy choice; it's also a political, economic and environmental choice. When you look at it that way, kale becomes a lot sexier.

It's true: each of us votes multiple times a day for the food system we want. Agriculture has significant impacts on our environment, economy and health. If we want clean drinking water for our community, we can choose to buy from farms that don't spray chemicals and that plant cover crops to prevent runoff in the rainy season. If we want to support our local economy, we can commit to spending the majority of our grocery dollars on local food from local stores.

As customers, we can even use our purchasing power to force stores to offer local products, like the loyal and influential customers of the Speerville Flour Mill. The mill's devoted customers are one of the key reasons this mill in Atlantic Canada continues to hold its own as a financially stable business after more than 30 years.

If you believe you can't make a difference as just one person at the grocery store's check-out line, think again. The Speerville Flour Mill is a success story, even though its products are purchased by only one to two percent of the local population. That one to two percent makes enough of a difference to keep the mill, its employees and 25 organic farmers hard at work.

On the other hand, a decline in customers can turn an innovative project that made news across the country into a memory. While the Kootenay Grain CSA continues to grow and supply local grain, the lack of support from CSA members and volunteer organizers has made it a shadow of its original self.

We all eat, and therefore we all get a vote. What are you voting for today?

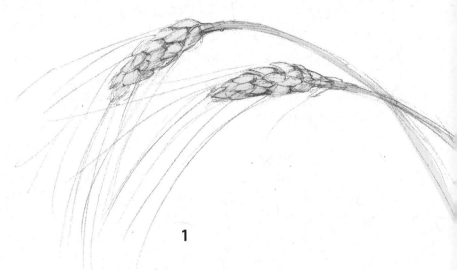

1

The Kootenay Grain CSA: Growing Grains to Leave a Smaller Carbon Footprint

*To produce our own food is the beginning of independence.
To accept that responsibility is the first step toward real freedom.*

— Gene Logsdon

THE KOOTENAY GRAIN CSA BEGAN WITH A BOOK: *The 100-Mile Diet: A Year of Local Eating* by Alisa Smith and James B. MacKinnon. Published in Canada in March 2007 and in the United States in April 2008 (under the title *Plenty: One Man, One Woman, and a Raucous Year of Eating Locally*), the book describes a young couple's year of eating only foods grown within 100 miles of their apartment in Vancouver, British Columbia, on Canada's West Coast.

Their 100-mile radius included the Lower Mainland, the southern half of Vancouver Island, the Salish Sea between the two areas and Washington State's Whatcom and Skagit Counties.

The book was timely, aligning with a growing interest in farmers markets and the locavore movement. (Barbara Kingsolver's book *Animal, Vegetable, Miracle: A Year of Food Life*, written with Steven L. Hopp and

Camille Kingsolver, was released shortly thereafter, with a similar theme of eating locally — but from a hobby farmer's perspective.)

Smith and MacKinnon's book inspired a swell of interest in local food, including "100-mile diet challenges" in communities across North America, with participants committing to eating only foods grown or produced within a set region.

Enter Matt Lowe.

Lowe, an environmental activist, was one of 150 Nelson, BC, residents who signed up for the area's "eat local" challenge in August of 2007. A member of the West Kootenay EcoSociety at the time, the environmental angle of eating locally appealed to him.

"I became interested in the idea of how we can be as sustainable as possible on a local level," he says. "This [eat local challenge] was manageable, as opposed to taking on all the problems in the world, which just seemed overwhelming."

Lowe committed to eating foods grown within 100 miles of Nelson for one day a week for a month. Like Smith and MacKinnon in Vancouver, he soon realized that his normal diet heavily depended on grains, and that locally grown grains were not at all easy to find. Lowe learned that most grains in his and his neighbors' cupboards had traveled an average of 930 miles.

Ironically, the fertile soils of the Creston Valley — only 77 miles away — had once produced award-winning grains. But many years ago, when

A Grain of History

During one of the CSA's farm tours, Tammy Hardwick, manager of the Creston Museum, told the story of Mrs. Amy Kelsey of Erickson, BC. She was the first woman ever to win the World Wheat King title in 1957. They were forced to change the victor's title to World Wheat Queen for the year. Much to the chagrin of organizers, Mrs. Kelsey went on to win again in 1958. She was banned from competition the following year when officials implemented a rule requiring a minimum growing area. It turns out that Mrs. Kelsey had been growing her award-worthy wheat in a small plot in her back garden.

the Canadian Wheat Board started setting quotas for their grain elevators, Creston Valley grain farmers found they were producing too much grain for the silos to take and so the surplus was left to rot in the fields.

As a result, by the 1970s farmers were growing other crops. By 2008, when Lowe had his sights set on local grain, he learned that the small amounts still being grown in the area were only for livestock feed.

Inspired and curious as to whether grains for human consumption could be grown in the region once again, Lowe contacted Brenda Bruns, a friend who lived in Creston and was involved in various community food initiatives.

Together, Bruns and Lowe surveyed some farmers in the Creston Valley and confirmed that there was indeed interest in growing grains again. Because of the area's mild climate, fertile soil and long frost-free

Fig. 1.1: *Kootenay Grain CSA shareholders touring Keith Huscroft's farm, one of the project's three growers, in 2008.* (Credit: Lorne Eckersley, *Creston Valley Advance*)

growing season, the Creston Valley is an agricultural paradise. They didn't think the actual growing of grains would be a problem.

All the pair needed were farmers willing to experiment and folks willing to buy local grain. Bruns and Lowe decided to bring these two groups together.

The first formal planning meeting was held in December 2007. Fourteen eager organizers from the town of Nelson and the nearby Creston Valley attended, including three farmers who were interested in being growers for the project.

"We were all coming to the meeting with the same good intention," Lowe recalls.

The committee decided to model the project as a community supported agriculture (CSA) program — and the Kootenay Grain CSA became the first grain CSA in Canada. With the CSA model, customers sign up and pay before a farmer plants his or her crops, so the farmer can match supply to meet demand. The up-front payment helps farmers pay their bills at a time of year when there is usually no product to sell and tangible income is often hard to come by. The funds are often used to purchase seeds and equipment, and invest in farm infrastructure such as fences and irrigation.

Then, during the course of the harvest season, members are rewarded for their early contributions with a share of the harvest as crops mature.

Risky Business

There are many factors in farming that are beyond a farmer's control, such as weather. And there are other risks that the farmer can try to mitigate, such as preventing pests and diseases by growing healthy plants and planting a variety of crops, so that if one fails, there is another to replace it.

The CSA model requires trust from its shareholders. They have to trust that the quality of the farmer's food will be worth their investment. It also requires some degree of confidence from the farmer. Farmers commit to a certain variety, quantity and quality of goods. If they don't follow through on that commitment, they will disappoint, and likely lose, their shareholders.

The community supported agriculture model is most commonly used to supply vegetables or fruits, but the CSA model can be used for many different food products, from honey or meat to eggs or seafood.

Depending on how a farmer structures their CSA, the risk can be shared between the farmer and the shareholders. For example, if it's a good year and the crops succeed, then the shareholders get a plethora of food. If it's a bad year, the shareholders walk away with less, just like the farmer. Some CSAs guarantee a certain variety or quantity of crops.

In 2007, the US Department of Agriculture estimated there were more than 12,500 CSA-supported farms in the United States, with another several hundred believed to be operating in Canada.

The Kootenay Grain CSA organizers believed the CSA model would ensure the farmers got a fair price for their product, time and labor, and would connect the farmers directly with those eating the grain. It was a win-win.

The committee determined they could offer 200 shares, with an estimated total of 100 pounds of various grains per share. They set the cost of each share at $100. If the yield was as expected, shareholders would pay $1 per pound for their organically grown whole grains. The price was less than the average commercial price of the combined grains. Once milled, the CSA's flour would cost substantially less than organic flour from a grocery store.

The project launched in early 2008, and with only minimal marketing, registrations in the CSA's first year came in quickly. All 200 shares were soon spoken for by 180 families and one Nelson bakery, Au Soleil Levant, which purchased 20 shares at a discounted rate of $75 per share.

The hard work had already begun as three farmers prepared their land to grow grains. The inaugural group included Ray Lawrence of Lawrence Farm, Keith Huscroft of Huscroft Farm and Drew and Joanne Gailius of Full Circle Farm.

Meanwhile, committee member Jon Steinman, a Nelson resident, began to tell the story of Canada's first grain CSA on his syndicated radio show, *Deconstructing Dinner*. His feature series, "The Local Grain Revolution," premiered on March 13, 2008, as the farmers were preparing their fields for the CSA's first season.

Fig. 1.2: *Keith Huscroft explains his operation to Kootenay Grain CSA members during a 2008 farm tour.* (CREDIT: LORNE ECKERSLEY, *CRESTON VALLEY ADVANCE*)

Thanks to local newspapers and Steinman's radio series, the public and mainstream media — including the *Globe and Mail* — began to pay attention to the Kootenay Grain CSA. Steinman's show began to inspire other communities to look at how they too could grow and eat local grains.

Back at home, residents of Nelson and the Creston Valley were buzzing with excitement — so much so that unexpected partners stepped forward, wanting to participate. The Kootenay Lake Sailing Association even proposed that the CSA take the environmental initiative one step further. The club offered to transport the bounty of harvested grain across

Fig. 1.3: *It was all hands on deck to help load the boats when mariners from the Kootenay Lake Sailing Association ferried the Kootenay Grain CSA's harvest across Kootenay Lake.* (Credit: Brian Lawrence, *Creston Valley Advance*)

Kootenay Lake by sailboat, thereby avoiding the fossil fuels needed to truck the grain an hour and a half around the lake.

Back in the three farmers' fields, the grain crops struggled through the dry, hot summer. Because the farmers had only the winter and early spring to prepare, their fields were not in tiptop shape for grain growing. As a result, some grains were planted later than what would have been ideal, and the farmers faced challenges with weeds and soil fertility. Nevertheless, the fields of khorasan, barley, wheat, oats and spelt grew — just not as well as the farmers had anticipated.

As harvest time grew closer, the organizing committee soon recognized their error in choosing to grow spelt and oats — two notoriously difficult grains to thresh without specialized equipment. The group added that to their to-do list, and by harvest day they'd tracked down the machinery needed to process the grains.

Once the wheat, spelt and oats were harvested, it turned out, because of the growing challenges, the crops hadn't produced as much grain as expected. Shareholders would receive a little less grain than anticipated. At $1.11 per pound, it worked out to 11 cents more than the estimated price — not too far off.

With large bags of whole grains about to arrive in members' kitchens, some of the expectant recipients began to ask how they would turn those grains into flour and food. Lowe, Steinman and their fellow organizers realized that many shareholders didn't know how to cook with whole grains, and didn't have access to a mill. Luckily, local expertise and equipment were available.

Creston-based Jennie Truscott and David Everest of Nelson both offered to mill grains for shareholders while Lorraine Carlstorm, a nutritional consultant, offered cooking classes featuring sprouting and sourdough.

With the first harvest over and the CSA's members gleefully experimenting with local grains in their kitchens, the CSA celebrated a successful first year. In September 2008, Lowe led the organization of the Kootenay Harvest Revival, a two-day celebration of local food to celebrate the CSA's success and raise awareness of the program. During the event, dozens of community members signed up for the 2009 CSA.

In spite of the challenges, Canada's first-ever grain CSA was a success.

That winter, the committee surveyed the CSA's shareholders and were pleased to learn that 83 percent of respondents had found the experience to be "definitely" fulfilling, while 15 percent had found the experience "moderately" fulfilling. (The remaining two percent were undecided.)

Most importantly, of 118 survey participants, 115 confirmed that they wanted to purchase a share in the CSA's second year. That was a true sign of success.

Excited by their promising first year, the Kootenay Grain CSA's organizing committee expanded its horizons. They set out to quadruple their crop production and triple their membership to a whopping 600 shares. It was a giant leap of faith and an equally massive undertaking, especially given that the group was volunteer-based at the time.

"We tried to grow a little bit too fast," Lowe admits in retrospect.

In addition to the previous year's crops, a one-acre test plot of lentils had been a success despite weed challenges, so it was added to the CSA in year two. Share prices went up. Each share would cost $125 and would consist of 20 pounds each of upwards of six various grains.

Once again filled to capacity, the CSA's second year began in earnest, and come harvest time, everyone was pleased to see bumper crops of wheat.

However, a lack of communication with regard to the health of some of the crops at Huscroft Farm and the subsequent poor yield left organizers scrambling to deal with the harvest's shortfalls and ensure CSA members received their fair shares.

Luckily, Lawrence and Gailius offered to pick up the slack with the surplus grain they'd produced, and shareholders weren't let down. But this decision left those two farmers with less grain to sell privately once their CSA obligations were met.

The issue marked just the start of a very difficult harvest season for the Kootenay Grain CSA.

Just a few weeks after the harvested grain was transported across Kootenay Lake by an even larger sailboat flotilla than in 2008, Lowe received an e-mail from the Canadian Food Inspection Agency with the

Fig. 1.4: *Farmer Keith Huscroft tosses a sack of grain to Drew Gailius at the Kuskanook Marina on Saturday, Oct. 17, 2009.* (CREDIT: CHRIS SHEPHERD, *KOOTENAY WEEKLY EXPRESS*)

subject line: "STOP LOCAL GRAIN MOVEMENT." In the message, the CFIA ordered the Kootenay Grain CSA to stop all shipments of grain to the East Kootenay and Boundary regions, where many of the shareholders lived. While these areas were not that far away from the Creston Valley where the grains were being grown, they were outside what the CFIA considered acceptable.

If the CSA wanted to transport grains to outside of the Creston Valley or Nelson, the CFIA said that the grains must first be tested for dwarf bunt fungus, which could be spread by planting contaminated seed.

The order frustrated the group. The CSA shareholders were buying grains to eat, not to plant, and so the risk of spreading dwarf bunt fungus was virtually nonexistent.

"We have offered to sign an affidavit that our grain will stay within a 100-mile radius and be used only for consumption, but so far the CFIA has declined our offer," Lowe told shareholders at the time.

With no other options, the group had to stop selling shares to willing members outside the CFIA's designated region.

Even more trouble was on the horizon.

Throughout the nearly two years of the program, trust in the CSA had never really wavered. The group navigated the ups and downs together. However, one incident put that trust in jeopardy and nearly sent the entire program into a tailspin.

Some may argue it did.

After distribution of the harvest, a shareholder reported something unusual about some of their wheat. It was wheat, sure, but it was pink.

It turned out that, unbeknownst to the group, some wheat from Huscroft Farm had been transported in a container that had once held grain seed treated with fungicide. Bad news — the grain was contaminated and could be unsafe to eat.

In a frenzy of activity, all wheat grown at Huscroft Farm was recalled. But the damage was done. The trust that the CSA so relied upon was broken.

It was a huge embarrassment, and the program's reputation as a source of safe organic food was seriously damaged.

Some shareholders accepted replacement grains, while others asked for their money back and left the program altogether.

It wasn't the way season two was supposed to end. What's more, volunteers were burnt out, and a tense wrap-up meeting led to Lowe's departure from the CSA organizing team.

That winter, Drew and Joanne Gailius left the program to start their own grain CSA.

Membership declined over the next three years, but the CSA was still operational in 2012 — albeit a shadow of its original self.

"It proved it could be done. We can do affordable small-scale grain production," Lowe says. "But [the CSA] no longer exists in its original form."

By 2012, the Kootenay Grain CSA was being administered not by the collective of eager volunteers that had started it all, but by the two remaining farmers: Lawrence and Huscroft.

Each shareholder signed up with a farmer directly, and while it kept the CSA alive, the community spirit the operation once relished was waning. That's not to say there wasn't still a demand for local grain.

"The farmers seem to be happy with the numbers," Jon Steinman says. "And, although they would welcome more members, they don't necessarily have the skills to really effectively market what they're doing."

As the calendar turned to 2013, Steinman was busy filming a new television series based on his popular *Deconstructing Dinner* radio show, but he still plays an integral role in keeping the project alive.

Steinman believes in the project so much that he's picked up a lot of the work the volunteer group used to do, from storing grain to milling it with a small commercial mill.

"Personally, I would have loved to see the momentum carry itself, but that didn't happen," he says. "I'm still very passionate about it, and I know there's a lot of people out there that are still passionate as well, but not necessarily in a position to actively go out and make it happen."

Steinman is in the midst of trying to gather enough people, "about 20 to 30 people to all go in on buying a nice substantial mill for the community."

It's clear that the success or failure of community grain projects such as that in the Kootenays depends on the support of its members and volunteers both as organizers and as consumers.

"When this project seemed to lose a bit of its momentum, it was because there wasn't any established organization to continue that momentum," Steinman says.

He's setting out to change that by encouraging the creation of such an organization.

"Every community is going to be different in terms of what type of infrastructure already exists — whether it's human resources or actual organizations or businesses or people that want to support these things," Steinman says.

You work with the strengths you have.

Steinman's radio show inspired a host of other communities to use their strengths and create their own local grain projects, including Urban Grains in Vancouver and Island Grains on Vancouver Island.

In fact, while the Kootenay Grain CSA has traveled a bumpy road, Steinman's coverage of Canada's first grain CSA may in fact be the project's most significant legacy.

Join the Revolution

- Sign up for or start your own "eat local" challenge. Try to eat local food for at least one day each week for a month. Spend some quality time on the Internet and figure out how far your food usually travels from where it's produced to your plate.
- Join a grain CSA. If there isn't one in your community, set up a meeting to gauge public interest in starting one. Make a special point of inviting farmers, millers, chefs and bakers to join the conversation.
- Consider the environmental impacts of how your food is produced. Do the farmers you support with your grocery dollars value the health of their employees, work to keep the community's water supply clean and consider themselves stewards of the land they work? Tour local farms and ask the farmers about their values and growing practices whenever you have the opportunity.
- Volunteer. Use your unique skills and resources to support a project you believe in. Anyone — from website designers to radio show hosts — can become an integral part of the community grain revolution.

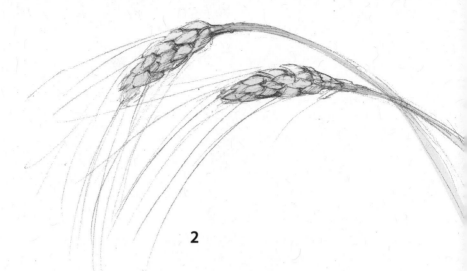

2

The Speerville Flour Mill:
A Community Flexes its Consumer Muscle

The wonderful thing about food is you get three votes a day.
Every one of them has the potential to change the world.

— Michael Pollan

IN 1973, A BRITISH CITIZEN BY THE NAME OF MURRAY HUBBARD decided to open a cooperative grist mill for local grain farmers in his community of Speerville, in the Saint John River Valley near Woodstock, New Brunswick.

The mill was shiny and new on the outside, but inside the equipment was mostly salvaged. The co-op purchased millstones from the abandoned Slater Mill in Kirkland, as well as some conveyors and other secondhand equipment in the area. Thanks to the co-op's thrift and determination early on, the mill was operational within a few years — but even then, only for the Speerville Flour Milling Cooperative's farmer members.

With only a few local grain farmers using the mill's equipment, it sat idle for long periods of time. That all changed in 1982 when a farmer in nearby Grand Falls asked the cooperative to purchase his wheat so it could be cleaned, milled and sold to a waiting buyer in Fredericton. While the cooperative had never before taken on this role of purchaser and distributor, the members agreed to the farmer's request. Co-op members Stu Fleishhacker and George Berthault made the 72-hour trek in treacherous winter conditions to pick up the wheat, process it and deliver the flour. In the end, the effort yielded more than just the milled grain. As a result of this experience, the co-op decided to expand its role beyond simply serving its member farmers and transform itself into a new company that would purchase, process and distribute locally grown grain. Speerville Flour Milling Cooperative Limited was born.

Over the next few decades, the mill established itself as the heart of Atlantic Canada's local grain community. It is still the only certified organic food-grade mill in Atlantic Canada, and as such has made it possible for Atlantic farmers to access the higher-value certified organic grain market.

The local community doesn't just benefit by having a source of local organic grain: the mill also employs an increasing number of staff. The mill welcomed its first full-time employee, Todd Grant, back in 1989. In 2004, Grant became the principal shareholder and president of the company. With his leadership, the Speerville Flour Mill has seen steady growth ever since.

Curiosity about the outfit's operations skyrocketed in 2009 as a result of the publication of — you guessed it — Smith and MacKinnon's *The 100-Mile Diet*. The book encouraged eaters to question where their food came from, and the mill had an answer. Their products were made from Atlantic-grown grains. But there was more to the mill's success than an increasing interest in local food. It was the mill's business policies that kept it afloat, according to mill spokesman Richard Wetmore, who joined the operation in 2007.

"We understand the capitalistic approach to business," Wetmore explains. "We understand fiscal responsibility very well. But we also feel that there's a responsibility to be a social entity. We're supposed to want to be big flour barons, but we just don't believe in that. Any chance we get to use our corporate power to help a community or a farmer, we will do that. Because we know that if we can help them, we will in turn help ourselves."

Over the years, more farmers began to see the benefits of working with the mill.

The mill has sourced all of its bulk grains from farmers in the Maritimes since 2005. Around 25 different farms work with the mill each year, and 12 to 16 of those farmers grow grain. In 2012, 14 farmers had a total of 1,145 acres of grain in production for the mill.

"We believe that, in order for us to have a successful value chain, we must have a successful farmer at one end and a satisfied customer at the other," Wetmore says. "If those things don't work, then that chain will break. So we do everything in our power to make sure that the organic farmer who grows the grain for us is successful."

The mill's goal is to provide a market for grain crops, like oats, that are usually grown as part of a crop rotation. These crops are grown to improve soil fertility. Being able to sell and profit from these crops is a perk.

"If a farmer has a 40-acre field that is in oats one year, we want them to be able to sell those oats to us so that block of land will realize some money," Wetmore says. Without a mill to process the grain crops, they

> "Any chance we get to use our corporate power to help a community or a farmer, we will do that. Because we know that if we can help them, we will in turn help ourselves."
> — Richard Wetmore, Speerville Flour Mill

Getting into the Spin of Things

Crop rotation is when farmers grow different kinds of crops sequentially on the same plot of soil. Crops are rotated in order to:

a) return nitrogen to the soil, by growing legumes such as peas or soybeans;

b) improve soil structure, by growing crops with deep roots (e.g., rye) and shallow roots (e.g., corn); and

c) prevent the buildup of pathogens and pests that is often seen when a single crop is continuously planted in the same area.

would be tilled back into the soil at a financial loss to the farmer. The Speerville Flour Mill's existence means that there's more food for the community and more profit for the local farmers.

In addition to oats, the mill buys spelt, rye, soft winter wheat and Red Fife wheat from farms in the Atlantic provinces. The grains are dried, after which samples are taken to test for disease, dryness and protein levels. In the cleaning stage, dead plant matter, weeds, stones and other non-grain materials (the chaff) are removed. Any disease-damaged or undersized seeds are removed to ensure the quality of the grain.

To make flour, whole grains are ground between stones. The resulting flour has a nuttier, sweeter taste and is more nutritious than the usual commercial flours, which are processed using steel rollers. While steel rollers grind grains into flour faster then stones can, the steel rollers remove the bran and germ (as well as the endosperm), where much of the grain's nutrition is concentrated.

Compared to stone mills, steel rollers also create more heat — which can destroy the vitamins in the grain. While stone-ground flour is often coarser, the reduced surface area lessens the flour particles' exposure to oxygen and subsequent nutrition loss.

Because the stone-milling process uses the entire grain, stone-ground flours have a shorter shelf life then steel-rolled flour — only six to nine months as opposed to years.

The Speerville Flour Mill does not use additives or preservatives, so fresh product is crucial: over time, the oils in the germ will go rancid. The best way to ensure fresh product? Sell it locally, of course.

Buying local grain also protects the mill from fluctuations in the world grain markets.

"People call us all the time and say, 'What are you going to do? The price of wheat just doubled.' It doesn't affect us at all," Wetmore says. "We don't give a crap what the world price is. We set the price for our grain at a point where the farmer can survive."

In 2012, the mill paid $500 a metric tonne (1.1 ton) for hullless oats and $600 a tonne for hard red spring

> "We don't give a crap what the world price is. We set the price for our grain at a point where the farmer can survive."
> — Richard Wetmore

Fig. 2.1: *This flour has a six-to-nine month life because there are no additives or preservatives.* (CREDIT: MAUREEN OVERMARS)

wheat: twice what oats were going for on the world market, and more than double the world price of wheat.

Wait, what? Why pay more when you don't have to?

"We know our farmers can't survive by growing wheat for $280 a tonne," Wetmore says. "We need farmers who can support us and support themselves."

In other words, the Speerville Flour Mill creates financially sustainable grain production. That's a big deal, and something that all local grain projects should strive for.

Paying farmers well has not harmed the business in the least.

"We do everything in our power to make that value chain viable all the way through," Wetmore says. "We make sure the farmer can make money, we do enough of a markup so we can afford to run our operation, and then we do our best to keep our prices competitive at the consumer end so the consumer can get a comparatively priced item but also a very healthy, local, organic food product."

For many of the same reasons as they have Maritime farmers as suppliers, the mill refuses to ship its final product very far either. They sell only within the bioregion of Atlantic Canada. "We believe that's the most efficient way of handling food," Wetmore says.

"Moving food 4,000 miles from where it's grown to where it's consumed is very inefficient. When we see California broccoli in local grocery stores that's cheaper than the broccoli grown in Nova Scotia, a caution flag goes up," he adds. "How is it possible for that broccoli to be grown in California, trucked all the way from California to Atlantic Canada, and be cheaper than what we can grow here?"

Wetmore believes it is questions like these that have more and more customers in his region buying the Speerville Flour Mill's products.

"They're starting to understand the gray areas in our food system," he says. "They're disheartened by it, and they'd rather support a system that's more sustainable."

It's that same community that's been the driving force behind the mill's ongoing success. But what if the whole community isn't on board with buying and eating local grain? Believe it or not, even a small percentage can make a real difference.

"Only one to two percent of the population buys from us, but that percentage loves our products," Wetmore says. "They are people who have gone out and asked questions about the food supply. They want to know where it comes from. They want to know the story behind it. These people are interested in supporting the things that we do. And that number has grown over the years to the point where we're able to have 13 full-time employees and pay all of our bills."

The mill's brand has become a staple for Atlantic consumers over the last three decades. Its loyal customers give the mill negotiating power when they talk to major grocery stores.

While large grocery stores usually prefer to have products warehoused, Speerville is instead able to negotiate regular direct deliveries to the grocery stores. This allows the mill to control product rotation and ensure that customers always find fresh products on the grocers' shelves.

"We know we don't have to bow to their wishes because the people who are buying our products in their stores are very faithful," Wetmore says.

If the mill's loyal followers can't get their products at a given store, those customers won't settle for anything else. Instead they'll simply refuse to shop there.

"And the local store managers love us, because they can order from us on a Monday, we can process the order on Tuesday, and they get the order on Wednesday. They love that it's fresh," he says. "They love that it's local, and the people that buy from them really respect that."

Over the years, that respect has translated into donations from members of the community and local businesses to help the mill purchase new equipment that allows it to introduce new products, such as oatmeal. Given its history, it should come as no surprise that the company often turns those donations around and gives back.

Community outreach became a focus in 2007 when Wetmore was hired to help educate the public about the benefits of healthy, organic local food. As a former teacher and garlic farmer, he is uniquely qualified to educate the community about the Speerville approach.

"When you're talking to the general public about issues with respect to farming and food, they don't have the same knowledge that farmers have," he says. "Our goal is to help them understand the importance of a local economy, of healthy food and of local organic food."

The Speerville Flour Mill works in partnership with a number of schools to provide healthy locally grown food for students and their families. In 2000, the mill partnered with Debec Elementary School in nearby Debec to introduce a hot breakfast program and provide pancakes and oatmeal made from Speerville Flour Mill grains for the students.

The mill also developed a fundraising program with its products so that instead of selling candy bars, students could sell local organic food to raise funds for school programs. The mill even offers a bulk-purchasing program for Atlantic schools that want to use the mill's products in their cafeterias. Shipping is free on 300-pound or $600 orders to anywhere in Atlantic Canada.

It's not just schools that the mill reaches out to. Since 2008, the Speerville Flour Mill has supported the local Healthy Families program for new parents. Wetmore speaks to participants about food labeling, nutrition and organic food. He makes healthy muffins or cookies using the mill's products to bring as samples — and he brings recipes.

"The point is to show people that it isn't that difficult to make your own healthy food," he says. "From an expense perspective, you will end up spending less money if you prepare it yourself."

Fig. 2.2: *The Speerville Flour Mill's portable wood-fired oven travels to community events across the Maritimes, including the Village Feast, an annual non-profit children's charity fundraiser in Souris, Prince Edward Island.* (CREDIT: LAURA HOGAN)

In addition to supplying customers through local grocery stores, the mill offers bulk-buying programs to families and food-buying groups. It also operates a store at the mill with a large selection of ethically produced products, such as peanut butter, oils, coffee and cleaners.

Farms and other small-scale businesses that want to use the mill's ingredients in their value-added products also qualify for the bulk-purchasing program. For example, bakers can use flour milled from locally grown wheat to make cookies that they then sell at the local farmers market.

Some farmers who make other value-added products use the mill as a distributor. Yum Bakery in Windsor, Nova Scotia, produces organic nut butters, while the Canadian Organic Maple Company in Divide, Nova Scotia, produces organic syrups and other maple products. The Boates family in Kentville makes organic apple cider vinegars. Because of the mill's relationships with grocery stores, these food producers can access the mill's distribution network and find their way onto grocery store shelves — which might otherwise be impossible.

In 1998, the mill partnered with local farmers to trial heritage wheats and grow the first commercial fields of Acadia wheat (an Atlantic heritage variety) since the 1950s. In 1999, the mill was awarded the Milton F. Gregg award by the Conservation Council of New Brunswick for its dedication to environmental protection.

After 30 years of steady growth, the mill is well positioned to take on any challenges that lie ahead. It has strong relationships with local farmers, a dedicated customer base and committed employees. Wetmore's only concern is being able to balance the supply of local grain with the increasing demand for the Speerville Flour Mill's products.

"Our fear is that the demand will increase so much that we're not able to supply the demand with grain grown in Atlantic Canada," he says. "We really do not want to have to mill grain that we buy from outside our area."

As long as the mill continues to have 12 to 16 farmers growing grains each year, the mill will be able to maintain its current sales levels and manage small annual increases. How's that for a promising outlook? And it's all because one day in 1972, Murray Hubbard decided Speerville needed a community grist mill.

Fig. 2.3: *Locally grown and milled flour is a source of pride in the Maritime provinces.*
(CREDIT: ACORN [ATLANTIC CANADIAN ORGANIC REGIONAL NETWORK])

Clearly, the Speerville Flour Mill's approach has worked. It is a model for mills like the Somerset Grist Mill in Skowhegan, Maine, and other communities that want to support local grain production. It proves that consumers can make a difference if they demand local products in stores.

Join the Revolution

- Flex your consumer muscles: shop with intention. Request that local stores carry local products and don't settle for non-local substitutes.

Grist for the Mill —
Knowledge is Power

IGNORANCE IS BLISS. It is a whole lot easier to never think about where our food comes from, to blindly shop at wherever is most convenient and to eat, without question, whatever the industrial food system provides.

Asking questions about where, exactly, our food comes from and how it's produced takes time. Many of us don't even know what questions to ask. We aren't the agrarians and homesteaders who settled North America, who knew how to milk a cow or plant corn by the time they were ten. We're not only ignorant of what farming used to look like — we don't even know what mainstream agriculture looks like today. Many North Americans still picture idyllic red barns and silos, cows grazing grass and rows of diverse vegetables being weeded by a grizzled farmer with a hoe. The truth is, the vast majority of food we eat is not from this imaginary farm.

Anyone who's seen Robert Kenner's documentary *Food Inc.* or read Michael Pollan's *The Omnivore's Dilemma* can tell you that the foods we consume today may be inexpensive and convenient, but they are often fraught with environmental, nutritional and ethical costs.

When we ask questions and educate ourselves about the real and potential consequences of mass food production, industrial food becomes harder to stomach. That being said, knowing more about our food, how it's grown and where it comes from also opens the door to new ways of looking at it. Education empowers us to make positive changes both in our diets and in our communities.

Those of us who want to learn more about our food can go beyond learning about feedlots, chicken factories and chemical (called "conventional" because it's the norm) vegetable production. We can learn a lot about our food by playing in our gardens. In fact, if you've lost your passion for food in the process of learning about industrial agriculture, growing edible plants is one of the best ways to rekindle your love affair with food.

Grains are a wonderful place to start.

Many of us think that growing our own wheat (or oats, or quinoa) is impossible unless we have a combine and a few hundred acres of land. But in truth, grain is one of the easiest food crops to grow, and it doesn't need much land to produce an impressive crop. The hardest part of growing grain is getting over the mental block so many of us have. Once you've accomplished the "impossible," once you've grown — and tasted — grain from your own garden, the rest of life's challenges feel a bit less overwhelming.

It was this awakening, and the desire to empower more people, that led Brock McLeod and his partner Heather Walker (now Heather McLeod, the co-author of this book) to host Island Grains — a community experiment in small-scale grain growing on their farm on British Columbia's Vancouver Island.

It's the sharing of this knowledge, through gatherings like the annual Kneading Conference in Skowhegan, Maine, that allows us to revive our local grain culture and tackle the next "impossible" tasks on our list. First: grow wheat. Then: change the world.

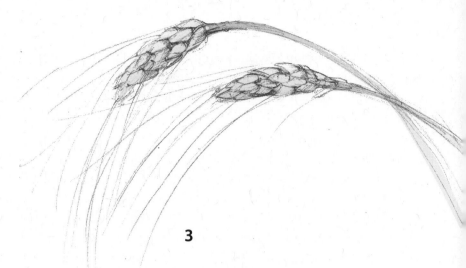

3

Island Grains: Learning and Experimenting as a Community

To be interested in food but not in food production is clearly absurd.

— Wendell Berry

IN 2008, BROCK MCLEOD AND HEATHER WALKER's goal was to turn their ten acres of pasture into a thriving vegetable farm.

Once vegetables were in the ground and a flock of hens were laying eggs, they considered other ways to bolster their self-sufficiency. They planted a nut and fruit orchard. They developed a long-term plan for honey, milk and meat.

But while they loved bread and dreamed of one day eating toast made with wheat they'd grown themselves, grain remained an elusive dream.

"I always thought that you needed acres and acres of land to grow grain," McLeod says.

After all, no one was growing grain in the Cowichan Valley. Maybe it was the small parcels of land. Or maybe the humid, damp climate made growing food-quality grain impossible. For whatever reason, most large farms in the region were dairy farms, and any large, open fields were being used to grow hay or corn for silage — not wheat.

With only ten acres and no combine, the very idea of homegrown grain seemed implausible.

Then McLeod stumbled upon a first-edition copy of Gene Logsdon's *Small-Scale Grain Raising* at a used book sale. According to Logsdon, McLeod didn't need a combine to grow grain. He didn't need hundreds of acres. He could grow the couple's year's supply of wheat on as little as ¹⁄₄₀ of an acre, or 1,100 square feet. At that stage of the fledgling farm's development, McLeod had plenty of space with which to experiment.

The door had been opened to a whole new world of possibilities.

McLeod discovered that a local seed company, Salt Spring Seeds, sold a wide variety of grain seed. He pored over the online catalog and picked a few to try.

In 2008, McLeod grew test plots of Red Fife wheat, khorasan wheat, oats and Faust barley to see how they would grow in his climate and clay-loam soil. He was pleasantly surprised by how healthy the plants were, and was taken by their beauty. Compared to his beds of strawberries and carrots, the rows of grain seemed exotic — and a little bit rebellious.

"It felt like we were doing something that we weren't supposed to be doing," McLeod says.

But even with this positive experience and all the excellent knowledge in Logsdon's book, he felt ill-equipped. He lived on the West Coast of British Columbia, not in Logsdon's Ohio. The winters are wet and the air humid, unlike the arid Prairie provinces where most Canadian grain is grown. Could the young couple grow enough food-quality grains on Vancouver Island to feed themselves?

Luckily, McLeod already knew many of the people who could help answer his questions, such as Dan Jason, the owner of Salt Spring Seeds, just a half hour from his farm. He could call them up and get answers, but he wondered if there were other people out there who had the same questions.

McLeod was seeing an increasing interest in local food: maybe there were other people on Vancouver Island who were interested in growing and eating their own grains. In *The 100-Mile Diet*, grain was one of the more impossible food items to hunt down — and the Cowichan Valley was within the Vancouver authors' 100-mile radius.

McLeod had listened to Jon Steinman's "The Local Grain Revolution" radio series, and was surprised to hear how excited the Kootenay Grain CSA participants were to visit the farms growing their grain. He began to brainstorm possibilities.

Soon, his idea began to evolve from a desire to be self-sufficient to more of a community education project. After months of planning, Makaria Farm launched Island Grains on December 26, 2008.

"Island Grains is a new form of community-supported agriculture," McLeod and Walker wrote on their website at the time. "It is a club, a workshop, an experiment and a movement. It is a way for eaters to be growers, to face the same challenges as our farmers, to learn to fear the weather and to understand the value of food. It's a resource for families, individuals, groups, risk-takers and food enthusiasts."

Island Grains offered a series of workshops featuring local guest speakers who would teach McLeod, Walker and their fellow "grainies" how to grow, harvest and thresh grain on a small scale.

> "Island Grains is a new form of community-supported agriculture. It is a way for eaters to be growers, to face the same challenges as our farmers, to learn to fear the weather and to understand the value of food."
>
> — Brock McLeod,
> Island Grains founder

And, because the young couple believed in hands-on experience and weren't yet using their entire ten-acre farm, they offered those who signed up for the project 200-square-foot plots of tilled soil to plant whatever grains they chose.

"We didn't know which varieties were best to grow," McLeod says. "The mass experiment opened the door to group discovery."

McLeod dreamed of bringing in Gene Logsdon as a star speaker. However, while very supportive of the project, Logsdon (who was 77 years old at the time) was reluctant to travel all the way from Ohio to Vancouver Island.

That motivated McLeod to look around, and he was pleasantly surprised to discover an abundance of local expertise. Workshop speakers

Fig. 3.1: *Robert Giardino, the president of the Heritage Grains Foundation, was one of the first speakers to visit Makaria Farm for the Island Grains Project.*
(CREDIT: HEATHER MCLEOD)

were found within a 100-mile radius, including from other Vancouver Island communities, from nearby Salt Spring Island and from Vancouver.

Speakers included Tom Henry, the editor of *Small Farm Canada* magazine and one of the few farmers to grow wheat on Vancouver Island. He lived in Metchosin, less than an hour away from Makaria Farm.

Dan Jason of Salt Spring Seeds also agreed to speak. Not only had he sold McLeod his test seeds, he was also the founder of the Seed and Plant Sanctuary of Canada.

"This is a super-exciting thing for me," he said at the time. "I've been trying to promote grains for 22 years now, and all of a sudden it's exploding."

Other workshop speakers included Mike Doehnel, a farmer who was experimenting with malting barleys

> "This is a super-exciting thing for me. I've been trying to promote grains for 22 years now, and all of a sudden it's exploding."
> — Dan Jason, Salt Spring Seeds owner

Fig. 3.2: *Patti Talbot and Bruce Sawatzky delight in trying cooked emmer grains for the first time during on of the first Island Grains Workshops.* (Credit: Heather McLeod)

on Vancouver Island; Robert Giardino, the president of the Heritage Grains Foundation and Katharina Gustavs, the writer of a whole-grain cookbook, who would teach participants what to do with their harvested crops.

To cover the cost of the guest speakers and their administrative time, the couple charged $65 per membership. One membership could be shared by friends or family members.

One of the first people to sign up for Island Grains was Sarah Simpson, a reporter from the local *Cowichan Valley Citizen* newspaper (and this book's co-author). She intended to write a series of articles on her experience, and after her first column, the registrations poured in.

The project resonated with other reporters as well, and McLeod did interviews with CBC Radio, TheTyee.ca and *Monday Magazine*. Island Grains was mentioned on countless blogs and websites.

Help Close to Home

One of Simpson's articles resulted in the discovery of Cowichan gardener Helen Reid, who had been growing quinoa for her own use in her back garden for more than a decade. She seemed a natural fit and was excited to share her knowledge with the Island Grains community.

Reid says she started growing quinoa as an alternative to rice because rice needs a lot of water and you never know when sources will become scarce. While not technically a grain, quinoa is known as the "mother grain" — it's a complete protein, and Reid found she could get a much higher yield per square foot than she could with cereal grains.

Plus, "It's very, very amazingly easy to grow," Reid says.

Fig. 3.3: *Helen Reid was a source of information right in Makaria Farm's backyard.*
(Credit: Sarah Simpson)

"We'd hoped for 20 members to make the workshops worthwhile, but after four weeks we had to draw the line at 50 memberships and started taking a waitlist, which eventually had 59 names," McLeod says. "People liked the idea of being able to grow their own grain — of doing something impossible."

Like McLeod and Walker, many people thought it was impossible to grow grain in the Cowichan Valley. But maybe it wasn't impossible after all.

People began showing up at their door to donate rusty sickles and scythes they'd rescued from storage. Why were there so many antique grain harvesting tools hidden in Cowichan barns? Simpson's first column, published in February 2009, explained everything. Citing *Citizen* columnist and local historian T.W. Paterson, Simpson wrote that "one of the first commercial crops exported from Vancouver Island was, in fact, wheat."

It seems that the Hudson's Bay Company got into the grain export business after they settled their Fort Victoria base in 1843.

But the grain export business wasn't what McLeod was after. His goal for Island Grains was education and empowerment.

"We were interested in bringing people in the community interested in growing grains together to learn," he says.

With spring looming and their grain-growing area still in pasture, McLeod broke ground for the grain plots in February, as soon as the field was dry enough to support his walk-behind rototiller.

He worked the soil again every few weeks, and soon it was time for the Island Grains workshops to begin.

Through the workshops, participants learned the basics of growing grains on a small scale. They learned which grains were easiest to thresh by hand and how different grains could be eaten. They learned how to prepare plots for sowing and how to efficiently plant their grain on a garden scale.

On planting day, everyone shared seed and the rakes they'd brought from home. Participants planted an amazing diversity of grains, from wheat and oats to chia and flax. It was the largest, and most disorganized, grain trial the Cowichan had ever seen.

Over the next few months under a hot blue sky, the chia withered while the lentils became a lush green foam. Flax blossomed into blue flowers, then became jangling gold seed balls.

Participants were encouraged to discover the life cycle of their grains by visiting Makaria Farm frequently throughout the growing season.

"We would often see little groups being led by a proud guide around the bronze stalks; many took home bouquets of their grain to dry," McLeod says.

The collective experiment proved educational in many ways.

Fig. 3.4: *Excitement spills over the fields during the first-ever Island Grains Planting day.*
(CREDIT: HEATHER MCLEOD)

While the rye surpassed expectations and towered beyond McLeod's reach, the quinoa became infested with tiny green worms and the seeds crumbled into dust. The emmer did well at first, then disappeared: it may have been a winter variety, unsuited for spring planting.

Because it was meant to be a learning experience, each participant was responsible for managing their own plot of grain. Ceding responsibility for other peoples' grains came with a significant downside, however. McLeod watched helplessly as many of the plots became overgrown with weeds that would, if not removed, produce seed and infest his fields for years to come. McLeod and Walker sent friendly reminders to the grainies, urging them to weed their plots, but ultimately the consequences of not weeding were part of the lesson they were learning. McLeod was learning a lesson too: there are risks to sharing your land.

At the end of August 2009, Makaria Farm invited all 50 Island Grains members and guests to attend what they dubbed Threshfest. McLeod taught harvesting, threshing and winnowing workshops in the morning

Fig. 3.5: *Island Grains members converge at Makaria Farm to harvest their grains.*
(CREDIT: SARAH SIMPSON)

and afternoon, and everyone paused for a potluck lunch. Many of the dishes featured quinoa, wheat berries and other grains. People harvested throughout the hot sunny day.

"While many plots had become overgrown with weeds, we were impressed by the bulging bags and bundles of grain that left our farm that day," McLeod says.

While the grainies had learned about the many ways to use whole grains in their kitchens thanks to the "Cooking with Whole Grains" workshop, milling their harvested grains into flour was still very much of interest to the group.

They were in luck. In Cowichan Bay, only ten minutes from the farm, was True Grain Bread, an organic bakery with a stone mill. Owner Bruce Stewart offered Island Grains the use of his mill. On Milling Day, a few months after the harvest, grainies came to get their wheat and rye ground into whole-grain flour. Island Grains members waited for their turn with the mill, watching the milling and eating treats from the bakery.

"That first year of Island Grains accomplished exactly what we'd set out to do, and more," McLeod says. "The goal was to learn how to grow grains on a small scale, and to engage with our community. But the ripple effects went well beyond Island Grains."

Simpson's nine-part newspaper series sparked community interest in local grains. A lot of people started talking about — and growing — grain. Her efforts were recognized in 2010 when she won multiple industry awards for community and environmental reporting.

The publicity also launched Makaria Farm into the local spotlight. McLeod and Walker won the Young Entrepreneurs of the Year Award from the local chamber of commerce.

"The resulting local media coverage was the best marketing a new business and farm could hope for," McLeod says. "It established our farm as an innovative, community-minded business."

That winter McLeod pondered what the second year of Island Grains would look like.

Despite its year-one success, some things would have to change. The major downside to the model was having participants be responsible for their own weeding. As McLeod feared, many of the plots had become overgrown, and the weeds had gone to seed. It would take years for the farm to recover.

Weedy grain plots also meant less grain for the project's members to harvest, and that irked McLeod.

North America's First Cittaslow Community

Thanks to the efforts of True Grain Bread's Bruce Stewart and other Cowichan leaders, Cowichan Bay was the first community in North America to earn the Cittaslow, or "slow city," designation in 2009.

Cittaslow is an international network of communities that value community relationships and quality of life. Rather than striving for development, Cittaslow communities seek to preserve the qualities that make them unique, including their history. They celebrate local foods and products.

"We wanted the experience to inspire our participants, not discourage them, and this meant they should end up with grain they could use in their own kitchens," McLeod says.

With these goals in mind, he revamped Island Grains for its second year.

"We offered the workshops again because we felt they were the heart of the experience: to empower people to be able to grow their own food," McLeod says. "But instead of participants having plots to experiment on, we would grow the grain for everyone."

That way, McLeod had more control of the soil, seeding, weeding and watering. Everyone would help with the harvesting, because that part was more than McLeod could handle himself without the proper machinery.

Taking a page from the Kootenay Grain CSA's book, McLeod decided to turn Island Grains into a grain CSA, with workshops included in the experience.

"We felt comfortable estimating a harvest that would give 20 shareholders each around 40 pounds of grain, and if the harvest failed completely, we would refund the payment, after deducting the cost of the workshops," he explains. "Refunding in the event of a crop failure isn't typical of a CSA arrangement, but it felt more comfortable to us."

Seedy Saturday

With education and empowerment as their goal, McLeod and Walker took the lessons they'd learned during the first season of Island Grains and offered their first "Grow Your Own Pancakes" workshop at Seedy Saturday in Victoria in 2010.

"It was standing room only," McLeod says. "Our goal was to share everything we'd learned and inspire more experimentation on the Islands."

The first Seedy Saturday event was held in Vancouver, B.C., in 1990 and was organized by Victoria's Sharon Rempel. Over the years since then, annual Seedy Saturday and Sunday events have spread to communities across the country and beyond. People come to talk about seeds, to trade and/or purchase seeds and plants and to attend workshops.

Makaria Farm launched the Island's first grain CSA in January 2010. The $200 membership included five workshops and an estimated 40 pounds of wheat and rye.

The new model attracted 17 subscribers — less than the 20 the couple had room for. "We were surprised that the CSA didn't fill to capacity, but it made sense," McLeod says. "Our first-year participants had learned to grow their own grains. Some were now growing it in their own gardens. Others just wanted to learn the skills, and they'd already done that."

The second season of Island Grains resulted in a successful harvest of Red Fife wheat, rye and hard white spring wheat.

To harvest the half-acre of grain, Makaria Farm invited Island Grains' members and friends to join in a work party. They harvested with sickles, scythes, scissors and pruning shears, and loaded the grain stalks into a convoy of pickup trucks (and one grain-stuffed car) to be fed into a farmer friend's parked combine across town.

Fig. 3.6: *The owner of this hatchback is still finding awns in the vehicle's nooks and crannies.* (Credit: Heather McLeod)

Each member received 37 pounds of grain, just slightly less than the 40 pounds McLeod had estimated.

After a successful second year, albeit significantly different than the first, McLeod yet again spent the winter pondering the future of Island Grains. He and Walker had achieved what they'd set out to do: they'd learned to grow their own wheat, and even had a generous supply in storage.

"Our cupboards were full of wheat and rye," McLeod says. "And our vegetable farm was expanding exponentially. We didn't have more room for grain."

Maybe it was the media coverage of Island Grains, or the popularity of Smith and McKinnon's *The 100-Mile Diet*. Perhaps it was the general movement toward local food sustainability that was growing in the province. Whatever the reason, people all over Vancouver Island and the Gulf Islands were feeling empowered to grow their own grains. And they were actually doing it — in both backyards and farm fields.

The couple decided that, while they no longer had the time or the land to support local grain production, they could still share their experiences and the information they'd gleaned from two years of workshops.

In 2011, Island Grains went digital, morphing into an online hub for grain growers to share their knowledge.

Join the Revolution

- Educate yourself. Plant some grain seeds and watch them grow.
- Host a potluck and challenge your guests to bring a dish that has grain as an ingredient. Give bonus points for those who use locally grown grain.
- Organize a farm tour and get to know your area's growers.
- Dig a little into the history of grain in your community. Was it being grown at one time? Is it still?
- Attend a Seedy Saturday event. If there isn't one near you, start one in your community.

Looking back at Island Grains' two years of success, McLeod thinks it was the project's timing that was key.

"The food security and local food movements were just beginning," he says. "Acting on our own interest in growing grain, we struck the right note at the right time."

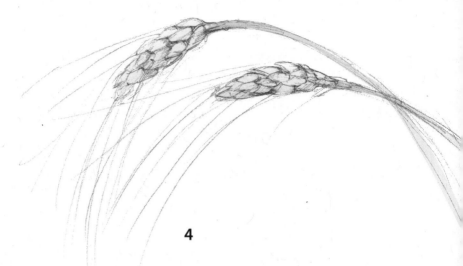

4

The Kneading Conference:
Connections and Cluster Economies

*The more we talk about local grain, the more people
and ideas can be included.*

— Amber Lambke

S INCE 2007, THE ANNUAL KNEADING CONFERENCE has brought bakers, farmers, millers, researchers and food lovers alike to Skowhegan,
Maine, a town of just under 9,000 residents. They gather for two full days
of presentations, panel discussions and hands-on workshops in the art of
grain and bread production.

While it does sound like a foodie's heaven, the conference's main purpose is more practical than just stuffing yourself with tasty baked goods.
Its goal is to connect stakeholders in order to revive the region's lost grain
economy, and by all accounts, the conference is doing just that.

Skowhegan is surrounded by fertile agricultural land. At one time,
the region grew all of its own wheat. In fact, according to Kneading
Conference co-founder Amber Lambke, "Not only did Maine grow an
abundance of wheat historically, peaking in the 1830s and '40s, but two

of grain harvesting's revolutionary inventions, the reaper binder and the horse-powered threshing machine, were invented in Winthrop, Maine."

Who knew?

By the 1860s, however, things were changing. Blame the trains. Grain could be brought in easily and cheaply from the Midwest thanks to the emergence of the transcontinental railways. So local farmers turned their attention to producing other crops. By 1930, wheat was no longer being grown in Maine.

Generations later, in the fall of 2006, Albie Barden returned home to Skowhegan from Camp Bread, a conference organized by the Bread Bakers Guild of America in San Francisco. Camp Bread focused on bread baking and the construction and use of wood-fired ovens. That year it had filled to capacity and turned away hundreds of other wannabe participants.

Barden owned Maine Wood Heat, a family business that built wood-fired ovens and taught others how to build and use them. He'd installed ovens in bakeries across the country, and was noticing an increasing interest among bakers in locally grown grain. He found it ironic that he was teaching other communities all around the world to build ovens, yet hadn't done so in his own community of Skowhegan.

After witnessing the success of Camp Bread, Barden wondered if Skowhegan could host a similar event focused on wood-fired ovens and artisan bread.

Early in 2007, Barden showed up at Lambke's door. He shared his idea. Lambke, a family friend of Barden's, was a speech therapist, but in her spare time she was just beginning to become involved in the economic revitalization of the community's downtown core, which was riddled with empty storefronts.

It was no secret the community needed a boost.

For a long time, Skowhegan had been a manufacturing town. Local factories produced shoes, toothpicks and other products, but beginning in 1950, many of those mills shut down and Skowhegan found itself in desperate need of new life and of economic development. It was so bad that, by 2007, around half of Skowhegan's 8,500 residents were eligible for food stamps.

Through her volunteer work in economic development, Lambke had been introduced to the idea of "clusters," or interconnected businesses within a given sector and geographical region.

Grain, for example, requires someone to grow it, harvest it, process it, distribute it, store it and turn it into food. When a community produces grain, it opens the door to any number of spin-off businesses and services from oven masons to brewers, from millers to pasta makers and so on. That's a cluster.

A 2006 report from the Brookings Institution called *Charting Maine's Future* had not only focused on the value of clusters, but also discussed the importance of agriculture in general. That made perfect sense to Lambke.

"Not only because people care about good food," she explains, "but because an agricultural economy preserves the open farmland that people come to Maine to see."

Lambke saw the cluster potential of grain. Maybe this would be the key to revitalizing Skowhegan's economy? So she and Barden brought together a group of professional bakers, wheat breeders, a miller, farmers, local business people and other stakeholders to brainstorm how they could combine the idea of a conference, economic development and local grain.

The organizing committee soon realized they couldn't just go to local farmers and ask them to grow grain as farmers had done so many generations ago. The knowledge of how to grow grain had been lost, the infrastructure had disappeared, and there was no market for locally grown grain.

The committee realized they'd have to start from scratch. They decided to begin by following the Camp Bread model and hosting a conference that would bring together stakeholders to focus on reviving the local grain economy.

They brought in non-profit partners to help organize the event, and six months later, in July 2007, Skowhegan welcomed participants to the first Kneading Conference.

Now held annually each July, the conference continues to grow and evolve. International interest in local grain has helped spread the word about the event, which offers hands-on workshops, lectures, demonstrations,

Fig. 4.1: *With so much to see and do, not a minute is wasted during the annual Kneading Conference. Handy chalkboard schedules help to guide the way.*
(CREDIT: JESSE COTTINGHAM)

panel discussions and information booths on growing, processing and baking with grain.

The registration fee includes all meals, many of which are prepared by participants in the workshops and includes the likes of melt-in-your-mouth-make-your-knees-buckle-it's-so-good pizza from wood-fired ovens, as well as pastries and bread.

"Our original vision started with farmers, millers and bakers," Lambke says. "Continuing the conference in consecutive years has allowed us to build the community of people across the country and world that are passionate about these issues and doing work on grains."

Before they knew it, the cluster began to grow.

"We began getting letters from an 80-year-old stone mill dresser. That was an eye opener," Lambke says. "There were people who dressed stone mills for a living. So we've brought those people into the conference."

Next came the brewers.

Fig. 4.2: *The only thing better than pizza is pizza from a wood-fired oven. Each year, participants of the Kneading Conference get to see what all the buzz is about.*
(Credit: Jesse Cottingham)

"The brewers were interested in what we were doing because it meant they could use local grains in their distilleries or micro-malting facilities," Lambke explains.

In 2011, the conference highlighted native corn traditions; in 2012, it featured rice growing.

"The more we talk about local grain, the more people and ideas can be included," Lambke says.

For Wendy Hebb, who attended her first Kneading Conference as a member of the press in 2008, it was life altering. She was hooked, and has worked as the conference coordinator since 2009.

"The overwhelming bad news of the world is somehow put on hold when you're in a community like this," Hebb relates. "These people are so down to earth. It gives me a great sense of relief and hope and encouragement. I like being around these people. They're not being stopped by anything. They have an idea, and they're just doing it, step by step."

Lambke says the conference is successful because it strikes people as authentic.

"They leave feeling like they've formed new friendships and connections in an informal atmosphere," she says.

A belly full of artisan bread doesn't hurt the cause either.

The conference caps attendance at 250 people, including volunteers and presenters, which fosters the intimate feeling for which it has become known. It has sold out every year. In the last few years, participants have started visiting the area early to explore Portland (Maine's capital city) and also the Kennebec Valley — which has resulted in tourism dollars. In fact, the majority of the conference's attendees come to the community from elsewhere, including Canada and even Europe.

"The idea is to revive regional grain economies no matter where you are," Lambke says. "People take this knowledge home, and they start

Fig. 4.3: *Nothing fosters friendships — both old and new — like freshly made food. There's plenty of that to go around at the annual Kneading Conference.*
(Credit: Cara Kennedy)

new bakeries, they start to grow grains, they start grist mills. There are pockets of activity happening throughout the world: Canada, Belgium, France, Hawaii ...."

In contrast to the swell of interest in the conference from around the world, it took a little longer for the average Skowheganite to buy in.

"It wasn't necessarily something that local residents knew much about or, at first,

> "People take this knowledge home, and they start new bakeries, they start to grow grains, they start grist mills. There are pockets of activity happening throughout the world: Canada, Belgium, France, Hawaii ...."
>
> — Amber Lambke,
> Kneading Conference co-founder

Locking up a Mill Site

In 2009, Amber Lambke and Michael Scholz purchased Skowhegan's historic jail building — just across the street from the town's old mill. For the next three years, they worked to raise funds to start the Somerset Grist Mill, which opened in September 2012. Under the brand name Maine Grains, they process local grain

in an attempt to meet the growing demand for local flour. The goal is to process upwards of 600 tons per year.

The renovated building has become a community food hub. It is home to the local farmers market, a local food café and a food box distribution program. The building also includes a commercial kitchen, which other businesses and organizations can rent and use.

Fig. 4.4: *The finished product is a source of pride and the result of a lot of hard work.* (CREDIT: AMBER LAMBKE)

were even interested in," Hebb says. "We live in a community that is really struggling. It's not what you would ever consider a foodie epicenter. They may have even been a little skeptical about it."

But over the years, the Kneading Conference has become a source of pride. Local businesses are now getting involved. In 2012, the Town of Skowhegan became an official supporter. That was one of the most exciting developments to date for Hebb.

"It's all rhetoric until a community actually begins to coalesce around these ideas," she explains.

Maybe it's that local interest that's behind the ongoing improvement and expansion and subsequent popularity of the conference.

The addition of an annual Artisan Bread Fair in 2009 on the Saturday after the conference has attracted local residents in droves.

"We wanted the community to be aware of this idea of supporting local farmers and local bakers, and cultivating local food," Hebb says.

Fig. 4.5: *With attendance capped at 250, participants get up close and personal with speakers at the annual Kneading Conference. It creates an intimate atmosphere and fosters a sense of community.* (Credit: Jesse Cottingham)

Vendors at the free festival include artisan bakers, cheese makers and other local food producers. Attendees can sample pastries and artisan bread, chat with professional bakers, listen to live music and watch demonstrations. Baking equipment and books are available for sale. An estimated 3,000 people attended the Artisan Bread Fair in 2012, an increase of more than 500 people over 2011.

"There really was no place left to park in the parking lot," says Dusty Dowse, one of the fair's organizers. He expects the number to grow in future years.

The bread fair not only engages the local community — it also helps create awareness of local food and a demand for local grains, which is gradually creating a new market for local farmers and entrepreneurs.

With new sales markets opening and a growing demand for local grain, some Maine farmers are now growing grain — 82 years after it was last grown there. Six farmers supplied the Somerset Grist Mill with locally grown grain in 2012, and Lambke expected that number to double in 2013. It's a massive triumph, especially after only a few years of the Kneading Conference. The positive spin-offs continue.

For Bob and Mary Burr of Blue Ribbon Farm, who have farmed in nearby Mercer since 1976, the availability of local grain meant they could add fresh pasta under the name Pasta Fresca to their product offering. The pasta uses Maine-grown ingredients, including Blue Ribbon Farm vegetables and flour from the Somerset Grist Mill.

Local companies have also incorporated Maine-grown grains into product lines, including Borealis Breads, Spelt Right Baking, Peak Organic Brewery, Oak Pond Brewery, Barkwheats Dog Biscuits, Heiwa Tofu and GrandyOats Granola.

New England Distilling, based in Portland, Maine, now uses grains grown in New England for its whiskeys and gin. Three years after the first Kneading Conference, in May 2010, a cooperative of 12 organic dairy farmers opened the first exclusively organic feed mill in the state. Maine Organic Milling sells grain grown in Auburn, Maine.

Meanwhile, taking a page out of their ancestors' books, to meet the growing need for smaller-scale grain-processing equipment on local farms,

New England engineers are designing custom equipment such as a solar-powered, small-scale thresher-winnower and a small-scale portable grain dryer. If the reaper binder and the horse-powered threshing machine were invented in the region, then why not smaller, more specialized gear?

In 2011, the Kneading Conference's organizing committee received non-profit status as the Maine Grain Alliance, and the Quimby Family Foundation provided a generous donation to purchase a portable wood-fired oven. The oven allowed the Alliance to expand its activities to include year-round community and school educational workshops, and to raise funds by selling pizzas and other baked goods at festivals.

2011 was a good year all around as it also saw the inaugural Kneading Conference West event, a three-day conference held in September in Mount Vernon, Washington. The sister conference to its Maine counterpart was spearheaded by Dr. Stephen Jones, a frequent presenter at the Kneading Conference and the director of Washington State University's

Fig. 4.6: *Knowing your neighbor and kneading with your neighbor is commonplace at the annual Kneading Conferences in both Maine and Washington State.*
(Credit: Cara Kennedy)

Northwestern Washington Research and Extension Center. It includes malting workshops and a two-day professional baking course.

Echoing the success of 2008 back East, the first "West" conference was well received. "The bakers loved it." Wendy Hebb says. "They don't [often] take time to share experiences with their peers. Everywhere you turned, there was a professional baker who could answer questions. It was wonderful for everybody."

It turns out that the American Northwest is just a little bit ahead of Maine in terms of resurrecting local mills, growing local grain and having bakers who are looking for local grains.

"But what was happening in the Northwest was that there were pockets of effort that were not connected, and so when we bring them all together for the conference, suddenly there's a dynamic that occurs and the Oregon baker who buys wheat from Montana finds out the farmer 15 miles down the road is actually growing that same kind of grain and selling it to Asia. Why not sell it to his neighbor? That kind of connection is happening, and you start to feel that this is a community of people," Hebb explains. "Then you find out there's an underground movement of people who are trying to restore grain and bread making in France, and Scotland and Canada."

The Alliance has been careful to cultivate relationships slowly and has stayed open to fresh ideas in order to expand the circle. "So we're not preaching to the choir — so we're not preaching at all," Hebb says.

In 2012, the Alliance established an official advisory board.

"It's been a slow evolution of careful cultivation of relationships with all kinds of people," Hebb notes. The Alliance continues to consider how it can expand its impact and help more people benefit from its efforts.

That support system has proven invaluable in making the economic impacts Barden and Lambke first envisioned a reality.

"Now that the infrastructure is there, entrepreneurs are coming forward " Hebb says, citing the new local mill as a prime example.

With interest in local grain increasing, bakers are being challenged to incorporate local flours into their production. However, these flours are not as predictable or uniform as the industrial flours most bakers are used to.

Hebb points out that, due to the small-scale production of local grains, it's very expensive to experiment with these flours. In response to this challenge, one of the Alliance's advisors, baker Brock Owens, is creating the Maine Grain Association Grain and Bread Lab. The lab will offer workshops for farmers and bakers so they can experiment with local flours, find out the best formulas and applications and then share that information. Other bakers and local grain aficionados have started bread labs in California and Washington. Washington's efforts are being led by Dr. Jones.

Despite all of the attention directed toward grains, Hebb predicts that in the future the focus will be less on local grain and more on culture. To her, grain is a symbol of something larger.

"My automatic sense of food has been this global, mythologized, 'out there' thing," she says. "Every time I make something from what I bought from the farmer who grew it, I feel this shift to a sense of my own

Fig. 4.7: *Many hands make light work. Many hands can also create a small community oven. It's one of the many workshops that participants of the annual Kneading Conference can sign up for.* (CREDIT: JESSE COTTINGHAM)

community and what it's doing and what matters in it. I think that is a cultural shift from global attitude to local attitude. It's not rhetoric, it's actual."

Has the conference's apparent success actually benefited the local economy? Because, after all, that was Lambke's original intent. Indeed, with the conference's growth and its expansion to Washington State, with the increase in New England grain production and the numerous businesses that now use local grains in their value-added products, an impact has been made.

Jeff Hewett, Skowhegan's economic and community development director, agrees that the local economy has benefited: "Absolutely."

"It has been opening a lot more doors for us," he says. "They've taken something starting with a very small idea to begin with and really have grown it remarkably."

> "They've taken something starting with a very small idea to begin with and really have grown it remarkably."
> — Jeff Hewett, Skowhegan economic and community development director

"It's a tough time for the area right now with the economy the way it is," Hewett says. "I'm not saying we don't have a lot of other good things going, but [the Kneading Conference] hit into an area that really had been neglected for a number of years from the agricultural side of things. This is starting to help it kick the other way. It's going to take some time."

Join the Revolution
- Attend the Kneading Conference, or Kneading Conference West.
- Organize a local mini-conference or community artisan bread fair of your own. Invite local farmers, millers, bakers, brewers and eaters to connect, collaborate and, best of all, break bread together.

Grist for the Mill —
Community Sufficiency

W HEN IT COMES TO FOOD, the goal of a self-sufficient lifestyle is to
sit down to a meal made entirely of ingredients you've produced
yourself. While it's possible to be self-sufficient, it's not always easy or
practical. It can be difficult and time-consuming to do everything yourself.

Growing a backyard vegetable garden is hard enough. Upping the
ante to caring for livestock presents an intimidating learning curve and
requires more time than most of us can spare. Replacing sugar with honey
from your own beehives is possible, but beekeeping is a challenging skill
— one that's becoming even more challenging with the collapse of honeybee colonies in North America.

Given these realities, perhaps the best meal we can strive for is one
where we know the people behind all the ingredients. Here's a challenge.
Before digging into your next meal, try to name the sources of all the
ingredients: from the beehive keeper to the dairy farmer, from the salt
producer to the person who grew your tomatoes.

For some of us, that may be impossible. Imagine how satisfying it
would be to know exactly where your food comes from. After all, you are

what you eat, and if you don't know what you're eating, then how can you know who you are?

While self-sufficiency refers to the individual, community sufficiency is when much of what a group or neighborhood needs can be found within that group or neighborhood.

Bryce Wrigley, the founder of the Alaska Flour Company, sums it up like this: "I'm a big advocate of people taking care of themselves. Not everybody can raise grain; there'd be no sense in everybody raising grain and everybody buying a combine. But if communities can work together and source some of their food locally, and strengthen their local economies in the process, then I'm in favor of that."

What this means is that, in order to eat local grain or any other food, you don't necessarily need to grow it yourself; you can support someone else who does it for you. Sharing the workload makes it easier for everyone.

However, when we become too disconnected, and the people we're supporting get farther and farther from home, as in the case of industrial agriculture, the idea of community sufficiency becomes lost. People no longer know where their food is coming from.

Community sufficiency is never anonymous. It is an entirely different experience to buy 20 pounds of wheat from a farmer you know that was grown on fields you've walked, than it is to buy wheat grown some 1,500 miles away by farmers you don't know. Thanks to industrial agriculture, the latter is, sadly, a more common experience for many.

The desire for community sufficiency is what prompted the creation of the Alaska Flour Company.

Wanting a deeper connection with their growers is what also inspired Jonathan Stevens and Cheryl Maffei's quest for local grain for their bakery and, as part of their journey, to the creation of the Little Red Hen Project.

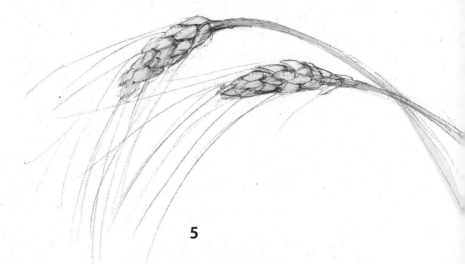

5

The Alaska Flour Company: How Hurricane Katrina Improved Food Security in the Last Frontier

Here's a state that is basically one-fifth the size of the United States with no flour mill any closer than 1,800 miles.

— Bryce Wrigley

FROM THE OTHER SIDE OF THE CONTINENT, livestock feed grower Bryce Wrigley and his family watched the devastation that followed the path of Hurricane Katrina in New Orleans and realized just how precarious their food security situation in Delta Junction, Alaska, really was.

"I remember seeing a picture of these people up on top of the roof and they didn't have any food. The reporter said that somebody had killed his neighbor for some food," Wrigley explains.

Despite New Orleans being only a hop, skip and a jump from the hub of some of the United States' biggest transportation and distribution networks, Wrigley watched authorities struggle for two weeks to get food to those in need.

"It just blew my mind," he says.

Having spent three decades up north, he's seen the import channels into Alaska shut down many a time. "Every three to four years, there's a transportation disruption that results in the shortage of food," he says. "This has happened pretty consistently for the 30 years I've been here."

In 1989, the Port of Anchorage froze, and upwards of 90 percent of Alaska's food supply vanished.

"The store shelves in Anchorage and Fairbanks and all around looked like a Third World country," he says.

Factors other than weather also affect Alaska's food supply. Dockworker strikes, hiccups in the supply chain and severe weather in the Lower 48 and elsewhere in the world also play a role.

"It doesn't have to happen in Alaska to affect us," says Wrigley. "We only grow about five percent of what we eat here: 95 percent is shipped in, so it really leaves us at the mercy of the transportation system."

And, as reliable as that system usually is, it is subject to breakdown and disasters, just like anything else.

The idea of increasing his community's food security had been rolling around in Wrigley's mind for several years, but seeing New Orleans reel from Katrina made him think more seriously about how his family and his farm could play a role.

"It seemed logical to us to begin putting in a local food system that would be able to take care of the basic nutrition needs for a period of time, just to weather these challenges as they come up — emergencies, disasters," Wrigley says.

His family has grown barley for animal feed for 30-odd years, but with Alaska's food security in mind, they began to see those fields of barley in a new light.

"What if we grew grain and milled it into flour? That's a basic commodity," he says. "We'd been trying to get somebody else in the state to grow it for quite a long time. There were people interested, but it just didn't seem like anything was happening."

While he was certain he could grow the grain, there was one thing standing in his way. There was no mill in Alaska.

"Here's a state that is basically one-fifth the size of the United States with no flour mill any closer than 1,800 miles," he says.

Fig. 5.1: *The move toward increasing their community's food security is a family affair for the Wrigleys in Delta Junction, Alaska. Bryce Wrigley is standing on the left.* (Credit: Bryce Wrigley)

Then, around the end of June 2011, the Wrigleys went on a family vacation. Their conversations were sprinkled with the idea of starting a mill, but this pipe dream became a reality after meeting an equipment supplier in Idaho. Wrigley decided to build a mill.

As for what he'd grow, the University of Alaska had just released a variety of hull-less barley that was suitable for human consumption. Rather than trying to grow wheat, which in the North would be a full-season crop and very challenging to grow consistently, the Wrigley family stuck to what they knew best: barley.

"We've never had a problem growing barley. It seemed logical, if we were going to do this for food security purposes, to focus on a grain that could be grown reliably," Wrigley says.

Because of the steep start-up costs, the family decided to start by building a small-scale mill to turn the barley into flour — a 30-inch stone mill, the same type of equipment used in a large operation.

"We built something we could add to in stages as our need grew, but that if it really went belly-up then we could pay it off ourselves and we wouldn't have any danger of losing the farm."

A few years later, the Wrigley family specializes in quality artisan, stone-ground flour. It is available in bags from 2 to 50 pounds.

Fig. 5.2: *It takes a keen eye and a little bit of skill to get the settings just right on his mill, but Bryce Wrigley is up for the challenge.* (CREDIT: BRYCE WRIGLEY)

A Pondering Physician

In the course of researching barley flour, a physician treating patients from Alaska's North Slope oil fields contacted Wrigley to inquire about his product.

"[He] was trying to get barley flour and cereal up on the Slope because they have so many workers up there who are coming in with stomach problems, and they were being diagnosed with gluten sensitivities."

Since barley is lower in gluten and higher in fiber than wheat, with a host of other health benefits, the idea was to give the workers the barley products to see if that would reduce the number of sick days they took.

"When it comes out of our mill, it looks like it does if you were to go to the store and buy it from somebody else," he says.

While initially there was no real community component to the project, slowly but surely the public is looking to join the revolution — some even going so far as to help grow grains for the Wrigleys to buy and mill.

"We are interested in working with other farms to grow for us," he says. "One of our goals was to create a new market for farmers here that wasn't in existence before." The mill is open to purchasing both barley and wheat.

Delta Junction residents have been very supportive as well. The small community's local store not only carries Alaska Flour Company's products, but ensures they're displayed nicely for all to see.

"It takes time to grow an industry. If it is to be available when we need it, we must support its growth," Wrigley explains. So far, that support looks promising.

Fig. 5.3: *There's something romantic and picturesque about working in the fields of Alaska, in the shadow of the snow-covered mountains.* (CREDIT: BRYCE WRIGLEY)

"We always are getting e-mails, and there's been a lot of support locally and around the state for what we're doing."

And really, he says, "That's what it's all about."

"I farm because I really enjoy it. I like to drive the tractor. I like to drive the combine. In a way it's kind of a selfish thing. But, when we started doing the flour mill, there was a new element," Wrigley says. "I still enjoy farming, but what we're doing now — by providing healthy local food for local people, there's a different level of satisfaction. We feel like we're doing something that is for the benefit of the local community, for the state of Alaska."

Join the Revolution

- Research how you would feed yourself and your family if your traditional food supply was cut off. Get prepared.
- Find out where your nearest flour mill is. Ask where they get their grain from.
- Increase community sufficiency in your area by buying local products.

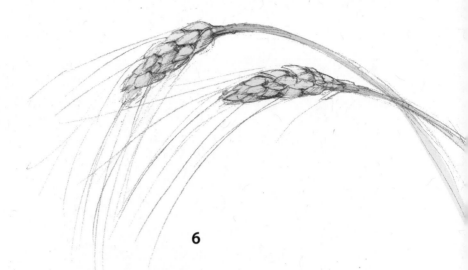

6

Hungry Ghost Bread's Little Red Hen Project: Getting to Know Your Growers

We could never have imagined this four or five years ago,
but we use local flour every day of the week.

— Jonathan Stevens

WHEN BAKERS JONATHAN STEVENS AND CHERYL MAFFEI opened Hungry Ghost Bread's retail shop in Northampton, Massachusetts, in April 2004, the duo was already thinking about sourcing local grain for their products.

They believed in the importance of knowing the people behind their ingredients, from the beets and carrots used in their baking to the wood they burned in their bread oven. Locally milled flour from Massachusetts-grown grains was a logical next step.

Logical, yes. Was it doable? Yes, but it would prove to be much harder than the couple anticipated.

A college town, Northampton is surrounded by some of the best soils on the continent. The nearby Connecticut River floodplains have rich silt soils, in which farmers grow crops such as potatoes, corn, other

Fig. 6.1: *The Hungry Ghost Bread bakers grind homegrown wheat with a hand-crank mill.* (Credit: Hungry Ghost Bread)

vegetables and tobacco. But where was the grain? It simply couldn't be found. Stevens and Maffei were surprised to learn that locally grown grain and locally milled flour did not exist in the region.

Reluctantly, they turned their sights farther afield and Hungry Ghost Bread started buying flour from Meunerie Milanaise in Milan, Quebec — a five-hour, 280-mile drive across the US-Canada border. The bakers were pleased to have a source of affordable, good-quality flour from grain growers they'd met within their corner of the continent, but they continued to dream of grain grown even closer to home.

If grain could be grown in Quebec, why not in Massachusetts?

The couple learned grain had once been grown in the area. Rumor had it that Massachusetts was actually the first place wheat was grown when settlers arrived on the continent. The Pilgrims grew grain in the region after their arrival in 1620. A water-powered mill had milled locally grown wheat and rye only a few generations ago in nearby Amherst.

The Hungry Ghost Bread bakers could confirm that grain had once been grown in the region — and therefore could be grown again. They had hope. But they were not farmers. They needed to find a local supplier.

Together with Leslie Cox, the farm manager at Hampshire College, Stevens and Maffei arranged a series of meetings with local farmers and bakers between 2004 and 2006 to discuss the feasibility of growing local grain.

"We thought it would be straightforward," Stevens admits. "It was not. It was a bumpy ride."

One major challenge was that farmers were reluctant to give up producing vegetables or other higher-value crops in order to experiment with planting grains.

"It's hard to sell farmers on growing something that we're only going to be paying a dollar-something a pound for, when they could be growing vegetables that are two dollars-plus per pound," Maffei says. "Grain was a hard sell."

The bakers tried to encourage the farmers to grow grain as part of their crop rotations, "Especially by smaller farmers," Stevens says. "If we can provide a market for the grain rotation, that's crucial."

But the Northampton farmers were too wary to take on the risk of growing grains. There were numerous challenges that would have to be worked out. Not only had they lost the knowledge of how to grow grain and which varieties grew well in the region, most farms did not have the appropriate infrastructure or machinery to grow grain on a commercial scale.

"They can learn how to plant. Planting something is not a big deal. Most farmers have the ability to clear a field. But — do they have a grain drill?" Maffei says. "They might have a corn drill that they can adapt. Do they have a combine? No. Do they have storage? Silos that are insect proof and vermin proof? No."

The devil's in the details, as they say. Despite these challenges, Stevens and Maffei continued to lobby local farmers to grow grain.

Finally, two local farmers agreed to grow trial crops of spelt, rye and some varieties of spring wheat. But the majority of the farmers declined. They were not ready to invest their time and money in finding the infrastructure, or in learning how to grow a crop that might not grow well or have enough of a local market to be profitable.

It was pretty much back to the drawing board when it came to sourcing enough local grain to supply the bakery's needs.

One day in 2008, Maffei, Stevens and Cox were brainstorming how their vision of locally grown grain could be realized. They understood that to grow the many, many tonnes of grains the bakery required was simply too much of a risk for a farmer to take on. There were still so many other questions: Which varieties would grow best in Northampton? How would they harvest, thresh, clean and mill the grain?

If the farmers weren't on board, where could they turn? To their customers, of course. Stevens believed the community would step up and create a demand, and that the farmers would see they had a built-in market to sell to.

For the past few decades, local vegetable producers had offered community supported agriculture programs, and they were thriving. Shoppers wanted local food. They understood the benefits of a community working together to support itself. Rather than self-sufficiency, Northampton's residents valued community sufficiency.

Fig. 6.2: *Nothing says bread festival like a marching band. Music is a staple for Hungry Ghost Bread's annual anniversary bash.* (CREDIT: HUNGRY GHOST BREAD)

"The customers — at least a certain demographic of food consumers — really get it here," Stevens says. "It was harder to convince farmers than it was our customers. Our customers were primed."

Hungry Ghost Bread had already helped prime the community to want local food. Each year, the bakers hosted a community celebration of Hungry Ghost Bread's anniversary with a Wonder Not! Artisan Bread Festival (because "you should know what's in the food you eat") — a community parade and celebration of music, poetry and local food.

A highlight of the September event is the sowing of grain seeds into what used to be a lawn in front of the bakery. Community members — mostly kids — broadcast the wheat or rye, rake it, stomp it and water it.

"We grow the patch of wheat at our bakery so when people come in they're looking at a wheat patch," Stevens says. "For many people, certainly for many children, that's the first time they've ever seen wheat. They had no idea where bread comes from."

Fig. 6.3: *Every spring, Stevens and Maffei plant wheat or rye in front of their bakery to remind their customers of where their bread comes from.* (CREDIT: HUNGRY GHOST BREAD)

This decision to tear up their lawn to plant wheat had been inspired by a book Stevens had read, called *The Bread Book* by Thom Leonard, co-founder of Wheatfields Bakery in Lawrence, Kansas.

While Leonard's idea of replacing lawns with grain was a bit crazy, it was also reasonable and that dovetailed with another of Stevens's inspirations — Abbie Hoffman, the leader of the Yippie movement in the '60s, who modeled a style of political and social activism that was "simultaneously outrageous and reasonable."

"It's a perfectly reasonable thing to do, to grow food," Stevens says.

And thus, in the spring of 2008, the Little Red Hen Project was born.

The Little Red Hen Project was named after the old folk tale and children's story about a hen that finds a kernel of wheat. None of the other farm animals will help her grow, harvest, thresh or mill the wheat, so the Little Red Hen does it all by herself — and then she savors the fresh-baked results of her labor while her jealous barn mates look on, mouths drooling.

Just as in the folk tale, the bakers asked: "Who's going to help us?"

It wasn't the farmers and their fields. It was the customers and their lawns.

Stevens and Maffei gave out three or four varieties of wheat grains to their customers, and invited them to plant the wheat seeds in their yards or gardens. More than 100 customers took home grain seed to plant.

What happened next was hard to believe.

"We put it out there, and lo and behold, we got more attention than we could ever have imagined," Stevens says. "The next minute, CNN is at my door at six in the morning, fitting me with an ear piece and asking me questions with a television camera in my face." The project was covered by *The New York Times* and featured on NPR.

In Northampton, the desire for bread made with local grain reached a near frenzy.

> "We put it out there, and lo and behold, we got more attention than we could ever have imagined. The next minute, CNN is at my door at six in the morning, fitting me with an ear piece and asking me questions with a television camera in my face."
>
> — Jonathan Stevens, Hungry Ghost Bread

Many Hands Make Light Work

When Hungry Ghost Bread built its first oven in 2004, a local kindergarten class came in and the mason read the story of the Little Red Hen to them. They went back to class with their art teacher and made tiles of the story. Those tiles are now installed on the bakery's oven.

Fig. 6.4: *Tiles depicting the story of the Little Red Hen surround Hungry Ghost Bread's oven.* (Credit: Hungry Ghost Bread)

"We were trying to grab some attention, and it worked, but it was way out of proportion," Stevens recalls. "This was at a time when wheat prices were rising astronomically."

Grain was making headlines around the world because of the fluctuating markets. The price the bakery paid for a 50-pound bag of organic white flour doubled between 2007 and 2008. The fact that the bakery was taking its grain production into its own hands was seen by many media outlets as newsworthy.

With grain growing in yards and gardens all over Northampton, and the project in the national media spotlight, the bakers' quest for local grain reached a tipping point. Somehow all that momentum and buzz the project received was enough to convince a handful of farmers to go ahead and grow some grain.

For a few years, five local farmers supplied the bakery. Eventually, infrastructure challenges proved too great for some of the farms. They couldn't get access to the right kind of machinery, and it was too time-consuming to clean the equipment after using it for animal feed in order to process food-grade grain.

But a few farmers soldiered on and continued to grow grains.

It turned out, after all was said and done, they really only needed one farmer willing to take on the risk. By 2012, Northfield's Four Star Farms was the local supplier for Hungry Ghost Bread, with up to 70 acres of grain in production. After years of persistence, Hungry Ghost Bread had its source of locally grown grain.

Farmer Gene L'Etoile had many reasons to start growing grain in addition to his crops of hops and sod. One reason was the desire to connect with the community in much the same way as his new baker friends.

"We wanted to have a crop that we could sell more to the local people," L'Etoile says.

Investing in grain production and milling equipment also helped the farm with succession planning.

"We have two sons who wanted to come back into farming, and we had to diversify to produce enough income for three families. There seemed to be promise in grains," L'Etoile says. While he believes the demand for locally grown grains is increasing, "We already grow more than we can sell, so it will take some increase in demand to absorb the current production."

Each week, the farm mills and delivers its wheat, spelt, buckwheat, triticale and corn flour to Hungry Ghost Bread. Stevens and Maffei are proud to use locally grown and milled flour and delighted with their ever-strengthening connection to their community.

"These people are our neighbor's," Stevens says. "Knowing who's involved with the process — it just makes sense."

All of the bakery's whole wheat and spelt flours are local. Between one third and one half of all flour used by the bakery is grown and milled in

Fig. 6.5: *Who says you can't have a little fun with your food? Not Jonathan Stevens.*
(Credit: Hungry Ghost Bread)

Fig. 6.6: *Fresh-baked treats made with locally grown and milled wheat are among the many rewards awaiting local grain revolutionaries.* (CREDIT: HUNGRY GHOST BREAD)

Massachusetts. (They still use white flour, khorasan and semolina from their friends in Quebec.)

"We could never have imagined this four or five years ago, but we use local flour every day of the week," he says.

With their supplier lined up for the foreseeable future, a new crop of challenges has emerged. But they are ones the bakery is happy to have — like learning to bake with flour other than ultra-processed commercial flour.

"There's got to be an educational process for bakers as well as customers. We're not all going to get those white, fluffy loaves. It's about difference. It's about variety in the gene pool, in the culture. If you're trying to get away from homogeneity, you gotta deal with variety," Stevens explains. "You can't have one without the other. We're not dealing with the Platonic ideal of what flour's supposed to be or what bread's supposed to be."

But teamwork often helps to solve those issues. Stevens emphasizes that the relationship between farmer, miller and baker is crucial for professional and home bakers alike — for anyone who is lucky enough to access locally grown grains.

"From year to year, the flour changes, even if it's the exact same variety of wheat," he says. "Different things happen in the fields. There's weather out there."

Stevens says that when he found one of the local flours he used was no longer performing in the way he needed, it wasn't that big of a deal. He was able to discuss the challenges with the miller (who also happened to be the farmer), and they were able to change how the grain was being milled. Four Star Farms was able to keep the bakery as a customer, which, in turn, could continue offering a locally grown product.

As for realizing their vision of baking with local grains, Stevens is proud of what they've accomplished.

"It's important for us to be part of our local economy," he says. "Keeping money in the community is the only way communities are going to survive. There is a system of commodity grains. If there's no counter, parallel economy, then that's all there's going to be. There's going to be privately owned seeds, commodity grain and commodity bread. Commodity bread will not feed us. It certainly won't feed our communities."

Join the Revolution

- Buy bread that's made with locally grown and/or milled wheat and celebrate the tasty inconsistency.
- Rip up your lawn and plant grain. At the very least, it will get your neighbors talking.

Grist for the Mill —
Making Connections

A GENERATION AGO, COMMUNITY MEMBERS CAME TOGETHER on a regular basis. For many, their gathering place was the local church or other place of worship.

That's not so much the case anymore. Many of the Baby Boomers' children have simply stopped going to church. Sunday football seems to have replaced Sunday services for some of us, along with the opportunity for like-minded people to gather on a regular basis for the greater good.

In many North American neighborhoods, adults live isolated lives with only a vague sense of real community — even when we're only an apartment wall or postage-stamp-sized yard away from another family. Many of those traditional connections our grandparents enjoyed and even relied on have been lost.

At the same time, we still long to connect, some of us more consciously than others. Humans are social creatures. It is our nature to want to be part of a community.

Perhaps that's why there are pockets of people who have sought out and found a way to recreate this sense of community that our grandparents enjoyed. These days, one of the ideas folks are rallying around is food. Connections are being re-established, not in a place of worship, but around the kitchen table.

People are realizing they don't have to look too far to form those connections, either.

Once there is a rallying point, such as Matt Lowe's idea of eating locally grown grain, or Island Grains' invitation to learn about and grow grains collectively, a community comes together.

As both the Kootenay Grain CSA and Island Grains show, expertise can be found right in our own neighborhoods. You don't need to talk to a wheat farmer from Kansas or Saskatchewan to learn how to grow grain. Someone locally could very well have been harvesting wheat quietly in their yard for years. Often, these experts are willing to share their knowledge or lend a hand because there is an inherent human need to use our unique skills to make the world a better place. While we may not all be experts at something, we do all have something to offer as parts of a greater whole, whether we're seed savers or masons or social media butterflies.

The Kneading Conference began in just this way, with a group of like-minded individuals choosing to gather for a specific cause and volunteering their unique skills and perspectives. The result was an annual conference that now brings grainies together from all over the world. The Kneading Conference's influence continues to spread when participants bring their knowledge back home and form new communities.

Give people a vision for change and a way for them to help make it a reality, and you will build community.

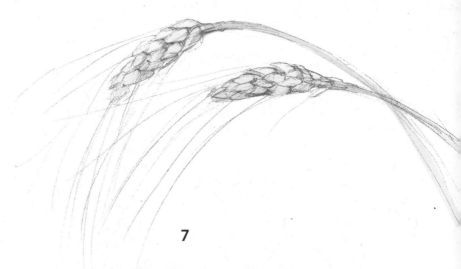

7

Chicken Bridge Bakery's Bread CSA:
Linking Loaves and Locavores

Without bread, all is misery.
— William Cobbett

IN 2008, AFTER WORKING AS BAKERS FOR A NUMBER OF YEARS at local bakeries in and around North Carolina's Chapel Hill and Durham communities, Rob Segovia-Welsh and his wife Monica decided they wanted more.

The couple was living in an old farmhouse in a rural part of the state when they came up with the idea to build their own wood-fired earthen oven and start baking from home.

"Originally we were focused primarily on supporting the farmers and brewers around us by using their products in our breads," Segovia-Welsh says. Potatoes, tomatoes, onions, cheeses, spent beer grains and the like "make natural and flavorful additions to any bread," he explains.

And lo, Chicken Bridge Bakery was born. Named after the Chatham County rural road where they built their first wood-fired oven, the bakery was originally meant as a side project. But it quickly escalated into Rob's full-time job.

Fig. 7.1: *It's hard not to smile when you have a bounty like this to share. Rob and Monica Segovia-Welsh are thrilled to be able to share their passion with their bread CSA members.* (Credit: Chicken Bridge Bakery)

"We are now a certified bakery with a large earthen oven," he says.

The bakers attend three farmers markets each week, and maintain a year-round community supported bread program.

Using the community supported agriculture model to sell vegetables is pretty common nowadays, and similarly styled grain programs are seeing a slow but steady increase as well. But community supported bread? Yes, indeed. A small number of these bread CSA programs are popping up throughout North America, from the Village Co-op's Kingston Bread CSA in Ontario to Good Companion Bakery's Bread CSA in Vergennes, Vermont.

In fact, bread CSAs may become more popular than grain CSAs because they take a lot of the hassle out of preparing grains. Instead of whole grains, or even a bag of flour, CSA members get a loaf of bread — often fresh out of the oven. Places like Chicken Bridge Bakery are going above and beyond by taking the value chain that one extra step and providing the final product. A loaf of bread can be more appealing than a bag of flour

Rob Segovia-Welsh of Chicken Bridge Bakery in North Carolina shows off his prized "breadvolutionary" loaf. (CREDIT: CHICKEN BRIDGE BAKERY)

A crowd gathers to welcome a flotilla of sailboats ferrying locally grown grain from the Creston Valley to Nelson, B.C.
(CREDIT: SAM VAN SCHIE, NELSON STAR)

Delta Junction, Alaska, farmer Bryce Wrigley harvests barley against an unlikely backdrop.
(CREDIT: THE ALASKA FLOUR COMPANY)

The tiles surrounding Hungry Ghost Bakery's oven in Northampton, Massachusetts tell the story of the Little Red Hen folk tale. (CREDIT: HUNGRY GHOST BREAD)

There are thousands of varieties of grain including these stunning wheats on display during an Island Grains workshop in Duncan, B.C. (CREDIT: SARAH SIMPSON)

Owners of the Hayden Flour Mill, the Zimmerman family not only mills the Heritage Grains Collaborative's wheat, but they also grow it for the Tempe, Arizona-based project. (CREDIT: HAYDEN FLOUR MILL)

Salt Spring Seeds owner Dan Jason demonstrates how to thresh grains using a homemade threshing box during an Island Grains workshop in Duncan, B.C.
(Credit: Sarah Simpson)

Chapalote flint corn is one of the oldest corns grown in North America and is making a comeback in southern Arizona thanks to the Heritage Grains Collaborative. (CREDIT: NATIVE SEEDS/ SEARCH)

Workshops at both Maine's Kneading Conference and Kneading Conference West in Washington include wood-fired oven construction and baking. (CREDIT: CARA KENNEDY)

Helen Reid shows that quinoa plants can flourish in backyard gardens on Vancouver Island in B.C. (Credit: Helen Reid)

Makaria Farm's Brock McLeod proves that all it takes is a little hard work and ingenuity to thresh grains on a small scale. (CREDIT: SARAH SIMPSON)

Fig. 7.2: *A Community Supported Bread (CSB) program lets Chicken Bridge Bakery's customers support the bakery at a time when it needs the financial boost. In return, they are rewarded with fresh-baked bread.* (Credit: Chicken Bridge Bakery)

to many who would like to support local business and eat better, but don't have the time to bake their own loaves.

When the Segovia-Welshes originally chose the CSA model, they did so for a variety of reasons. For one, they didn't have a retail shop.

"In order to make a living at baking, we quickly realized that we couldn't compete with large bakeries and their wholesale prices," Segovia-Welsh says. "We didn't have the ability to make huge quantities at a time anyway."

A CSA model allowed the artisans to continue to focus on making the product they wanted for the people who wanted it.

"We were able to connect with lots of local folks through this model, to the point where it doesn't feel like we're making a mass-produced product for nameless consumers," he says. "We get into discussions about topics they're excited about. They challenge us to make breads that they remember as children or from 'the old country.' It's a great exchange." Now that Chicken Bridge Bakery is a little more established, the CSA model isn't as integral to the operation as it once was. Retail sales at the farmers markets make up the bulk of the bakery's sales. But the pair won't part ways with their loyal CSA members because of the intimate connections they've made and continue to develop. At the same time, their bread CSA is still evolving. The bakery is collaborating with another local producer, Tova Boehm of Short Winter Soups, to put yet another spin on the community supported model.

Boehm offers a soup and salad CSA with the option of Chicken Bridge Bakery's breads.

"She serves 70 or so families and sells out within hours," Segovia-Welsh says.

The bakery also works with their friends at Small Potatoes Farm to offer bread as an option in the farm's weekly CSA program.

Fig. 7.3: *Being able to offer artisan bread with locally grown flour is something Rob Segovia-Welsh prides himself on.* (Credit: Chicken Bridge Bakery)

"For them, the inclusion of bread goes perfectly with their handmade cheeses, pasture-raised meats and fresh produce," Segovia-Welsh says, noting more options are always being bandied about. "There has been talk about creating a fermentation CSA that would include our breads along with locally produced kimchee, beer, yogurt, kombucha, tempeh and perhaps more."

With Chicken Bridge Bakery supporting local producers and small artisans, it means their bread CSA members do too. But getting local grains for their products has taken a little longer.

"The availability of locally grown, organic grains has really increased in the last few years here, and it's really exciting," Segovia-Welsh says. "We are just a small bakery and try to do our part to support the growers and millers who are really doing the legwork to make a wider variety of grains available to the area."

The bulk of the flour he and Monica bake with comes from family-owned and operated Lindley Mills in Graham, North Carolina.

"Although they supply us with hearty organic grains and flours from out west, they have also played a big role in encouraging growers in North Carolina to grow heritage breeds of wheat and are able to buy such quantities as to make it worth the growers' time and labor," Segovia-Welsh says.

Places like Looking Back Farm in Tyner, North Carolina, and Box Turtle Bakery in Chapel Hill have worked hard at seed saving and the propagation of rare heritage grain varieties.

In 2012, a mill called Carolina Ground opened up in Asheville, North Carolina, and is working hard to advance the local, organic grain movement there.

"They are contracting with local farmers and getting in touch with bakeries to get them on board with their freshly milled flours," he says. "We have been getting an old variety of rye [Wren's Abruzzi rye] from them, and the flavor and aroma of the flour is absolutely unique and exquisite."

Closer to home, Okfuskee Farm also produces Wren's Abruzzi rye so they are able to get the same rye berry from that farm — but not much, as the farmer only plants a small amount each year.

"I keep telling folks that it's an absolutely amazing time to be a baker," Segovia-Welsh says. "Heritage grains are making a comeback, there's a willingness and enthusiasm to support local food systems and there are great farmers markets that put the growers, artisans and customers all in one space."

Join the Revolution

- Join a bread CSA. If there isn't one, considering starting a bread club Chapter 12: bread club (see Chapter 17).
- Put out the call to those you know who make bread and get a sourdough starter. Sourdough starters often have amazing stories from being shared and passed down through the generations. Or, follow the steps in the "Eating Grains" section (Chapter 16) to make your own sourdough starter. Then learn to make sourdough bread.

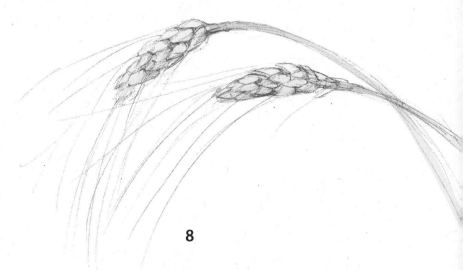

8

The Carbondale Community Oven: The Community that Bakes Together ...

The flavour of bread shared has no equal.

— Antoine de Saint-Exupery

LIKE MANY COMMUNITY GRAIN PROJECTS, the idea of a shared public oven is not new. Large communal ovens have been used as an efficient way to bake bread and feed civilizations all over the world, including in the Mediterranean, Europe and parts of Asia, from as early as 4000 BCE.

Families would prepare their dough at home and then bring it to the community oven to be baked.

Community ovens are an efficient way to use wood, because they eliminate the need for many smaller baking fires in individual households. For hundreds of years, these ovens produced very little waste, with the wood ash and breadcrumbs used as fertilizer and chicken feed. And since most villagers visited the local oven frequently each week, the oven became a natural hub. Neighbors discussed the local news and politics.

One thing is for sure: no matter what century it is, local news and politics never get old.

Maybe that's why communal ovens still survive in many parts of the world today. Except, that is, in North America, where most communities outside of Quebec (where French settlers continued the tradition) never had a community oven to begin with.

The majority of settlers who arrived in the New World didn't continue the practice of making dough at home and bringing it to a shared oven. Perhaps the abundance of wood in North America made it more convenient and efficient for families to have their own ovens. Or maybe settlers in rural areas lived too far apart for a communal oven to be practical?

While communal ovens aren't a part of most North Americans' history, there's been a recent surge of interest in this traditional practice. The last few decades have seen a move toward the establishment of community ovens in many North American towns. These ovens aren't meant to be a community's primary cooking oven, but rather a gathering point and community hub.

Fig. 8.1: *The beginnings of the Carbondale community oven.*
(Credit: Linda Romero Criswell)

The oven revolution is happening in communities clear across the continent, including in small towns like Carbondale, Colorado.

Sitting about 30 miles from Aspen in the heart of Colorado's Rocky Mountains, Carbondale is home to just over 6,400 residents. It's a unique town, proud of its community spirit and propensity to do things collaboratively. In 2009, that energy began to coalesce around the idea of building a community oven: a freestanding, wood-fired range in a public space that would be managed by a few, but accessible to everyone.

Over the past few decades, community ovens have been built in New York, Nova Scotia, Ontario, California and elsewhere. Residents come together to bake and eat together, and to share their stories. While the ovens themselves vary in design and construction, they are usually built in a park or other publicly accessible space. Community groups manage some, while others are run by local governments.

In Carbondale, the community oven project was spearheaded by resident Linda Romero Criswell, with the support of Tom Passavant, the co-president of Slow Food Roaring Fork, a chapter of Slow Food USA. The oven was built in 2010 at a public park next to a non-profit community center.

Why build a community oven? "[Human beings] would rather do things with each other than by ourselves," Criswell says. "Cooking can be quite a lonesome thing, and some of us thought it would be fun if we could bake together."

While the idea of a public oven likely makes liability-sensitive governments twitchy, the Carbondale oven organizers encountered minimal red tape in gaining permission to build their oven. Criswell says that the process of obtaining a building permit, doing the work, completing the inspection process, and getting the project signed off on took about a year.

"We have a typical town government that's pretty conservative," she says. "They don't want to get in trouble or anything, but there weren't too many hurdles." After all, says Criswell, "Who doesn't like hot homemade bread?" Public ovens are a magnet for community events, from fundraisers to festivals and school lessons. Unlike barbeques, grills or campfires,

Thinking Outside the Pizza Box

Community ovens can be revenue generators. For example, the ovens in Toronto's Dufferin Grove Park were built by volunteers, but are managed by the park's board and used to raise funds for the park.

Community groups and local residents can rent a public oven (with a trained staff member to operate it) for festivals, fundraisers and private events.

"We make a few dollars, and they either have their own chef or we provide somebody, and then they can say 'come to our spaghetti potluck with bread made in the community oven,'" Criswell says.

Many oven organizers, including Carbondale and Dufferin Grove Park, also host regular pizza lunches for the public. Park-goers bring their own pizza toppings and can buy dough, sauce and cheese for a few dollars. Nearby community gardens provide a handy source of fresh herbs, tomatoes and other toppings.

Fig. 8.2: *One of the first things in the community oven after it's lit is often pizza, a fan favourite.* (Credit: Linda Romero Criswell)

ovens are sheltered from wind, rain and snow, so they can be used all year. They can also have a secure, lockable door installed to prevent costly vandalism and accidents.

To date, the response to Carbondale's community oven has been encouraging. While the oven is large enough to bake about 18 to 20 pound-and-a-half loaves at once, the volunteers sometimes have to put in a second or third load in order to accommodate everyone who wants to participate.

Traditionally, unsweetened breads are baked when the oven is at its hottest temperature. As the oven cools, sweet breads, meat and casseroles can be cooked. Finally, at the end of its heat cycle, the oven's mild warmth can be used to dry herbs and fruit, make yogurt and dry wood for the next day's baking.

Pizza days are easier than bread bakes because the fire doesn't have to be started the day before.

Fig. 8.3: *Some modern bakers still practice the tradition of marking their loaves with certain cuts to identify them when they come out of the oven.*

(CREDIT: LINDA ROMERO CRISWELL)

"You just build a hot fire, and you're off to the races," Criswell says. Bread can then be baked on the following day, since the oven has already been lit.

Historically, each family would mark their bread loaves or baking tins with a unique design to distinguish their bread. Sometimes the order in which each family's bread was baked was organized by lottery, or the family names would be listed in order on a permanent plaque by the oven. But modern community oven groups often employ online sign-up forms, e-mails and various other methods to organize their baking times.

While it only takes about 40 minutes to bake a load of bread, lighting the fire, keeping it going and getting the oven to just the right temperature has become an art form. Somebody has to ensure there is enough wood

Fig. 8.4: *Linda Romero Criswell (left) and volunteer Katie Leonaitis break bread after the very first time the Carbondale community oven was used.* (Credit: Linda Romero Criswell)

for the duration and then start the fire on the afternoon of the day before the bread-baking day. Someone then has to check the oven late at night to make sure the fire keeps burning.

"It's definitely a team effort," Criswell says. "One person can do it, but it's a lot of work. If a lot of people do it, it's just a little bit of work for everybody."

Carbondale being a small town, it's not always easy to find enough volunteers to run a bake. But support for the Carbondale oven hasn't run short yet. The oven has been used every month in the two years since it was constructed, save for one month when a fire ban prohibited its use.

A core group of about 30 people is generally expected to show up and help, and with every use, that number seems to grow just a little bit. Criswell believes that trend will continue.

"We are located in a very vibrant place," Criswell says. The oven is in a park in the heart of Carbondale, next to a very active community centre. "We're going to start baking in conjunction with the events at the

Fig. 8.5: *A plaque on the Carbondale community oven's door tells it like it is.*
(Credit: Linda Romero Criswell)

non-profit centre, to draw the people that go to the gallery openings, the plays, the concerts and things like that. They're all interested."

With the future bright as more and more people learn of the virtues of community ovens, change in another aspect of the project is also underway. The Carbondale Community Oven group is excited at the potential to include locally grown grain in their breads in the near future.

An 80-year-old farmer whose land overlooks the park has stepped forward to supply the oven's users with the project's first locally grown grain. The rancher's mother used to grow wheat to feed their family, and after reading a magazine article on the oven project, he wanted to show his support.

"Wheat hasn't been grown in western Colorado for a long time, but he wanted to bring it back after about 65 years," Criswell says. The first crop of grain was planted in spring 2013.

With a farmer growing local grain and an oven to bake the bread, one thing is still missing: a mill big enough to clean and grind lots of wheat at once into flour. But Criswell is optimistic that the community will soon fill that gap.

"I predict that in about five to ten years someone will build a mill in this area and start growing wheat," she says. "I've had several farmers call me about it and want to do it."

In the early days of the community oven's planning, Criswell never expected these kind of connections to be made. "Most of what's happened

Community Oven Considerations

- Location. Find a public, visible space close to a water source, storage area (for wood, etc.), washrooms and, if possible, a community garden.
- Fuel. Wood-fired ovens need a supply of affordable wood, ideally hardwood.
- Support. Community members and groups will need to make use of the oven in order to justify the work of building and maintaining it. Letters of support, volunteers and donations will help get the oven built and operating.
- Management. Someone will need to organize the training and use of the oven.

has been totally unexpected," Criswell says. "It's exceeded my expectations. Here you have these old-time ranchers talking with these new-age, slow food yuppies, and we've really formed some interesting connections in the community. It's wonderful."

"Here you have these old-time ranchers talking with these new-age, slow food yuppies, and we've really formed some interesting connections in the community. It's wonderful."

— Linda Romero Criswell,
Carbondale Community Oven Project leader

Join the Revolution

- Rally your neighbors or community and build a community oven.
- Organize a public planning meeting to connect everyone interested in building the oven, and invite a guest speaker who can talk about how other communities have succeeded. Take the names and contact details of anyone interested in being involved: bakers who can teach workshops, construction crew workers, marketing gurus and more. Any subdivision can become a neighborhood if residents can gather for regular pizza parties.
- Ask your local library to stock books about wood-fired ovens and bread baking. Even better: buy them and donate them to your library.

Grist for the Mill —
Thinking Outside the Box

G RAIN, FLOUR AND BREAD ARE INNOCUOUS, everyday parts of life for most North Americans. We eat our boxed cereal or pre-sliced toast without asking what, exactly, we're putting into our mouths.

Once we learn more about these everyday foods — how they're produced and what's lost in the process — we might start to seek out alternatives. For some of us, that means buying food made with a specific variety of wheat (such as spelt) rather than the usual blend of modern wheats. Others might start to value homemade or artisan bread over the uniform loaves produced in factories. Some eaters might question the distance their grain has traveled from the field to their dining table (its "food miles"), and opt for grain grown closer to home.

So how do we help create and sustain these alternatives?

The 1960s troublemaker and social activist Abbie Hoffman preached taking actions that are "simultaneously outrageous and reasonable" to instigate change. That's the mantra the folks at the Little Red Hen Project followed when they ripped up their lawns to plant wheat. It also seems to be a running theme in the grain revolution.

A revolutionary must be willing to be outrageous and stand apart, because that's what it takes for people to take notice. The Kootenay Grain CSA got The *Globe and Mail's* attention when sailboats signed on to transport wheat in an environmentally friendly way. Outrageous? Yes. But more reasonable than trucking grain around a lake? Absolutely.

Island Grains triggered a frenzy of interest in local grain by daring to grow grain on the West Coast — something that wasn't supposed to be possible. But grain was one of Vancouver Island's biggest exports just a few generations ago. Outrageous? Yup. Reasonable? For sure.

Another project following Hoffman's "outrageous but reasonable" credo is that of Arizona's Heritage Grains Collaborative, which is growing very old varieties of grains with a seemingly outrageous twofold purpose: first, to combat poverty in Arizona, one of the USA's most poverty-stricken states, and second, to adjust to the effects of climate change by planting more drought-resistant crops.

Likewise, Nebraska-born farmers in California's wine country have started growing grains between the rows in local vineyards, as part of the Mendocino Grain Project. This may sound outrageous, but wine and bread are one of the most natural combinations in human history.

Sometimes thinking outside the box makes us realize just how unnatural that box really was.

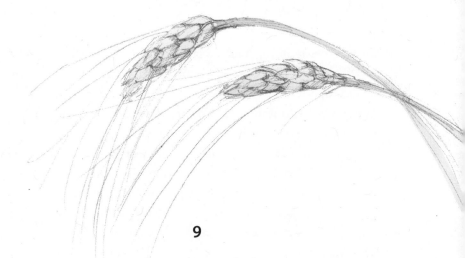

9

The Heritage Grains Collaborative: Tackling Poverty and Climate Change by Growing Grain in a Desert

*None of the heritage grain growers I know position themselves
as little Davids, as if they are out to slay the Goliath-like cereal
commodity corporations of the world. Instead, what they are offering
to us is taste, texture, nutrition and hope that can deeply change
our relationship to food.*

— Gary Nabhan

ACCORDING TO THE US CENSUS BUREAU, Arizona ranks among the ten poorest states in America. In 2011, the state tied with Alabama for seventh in the country, with 19 percent of its population living below the poverty line. One in four Arizona children lives in poverty.

The region's citizens face two major challenges: the struggling economy and finding a source of affordable, high-quality, nutritious food. In 2011, faced with poverty levels that hadn't improved over the years, a group of farmers, businesses and not-for-profit organizations in southern Arizona decided that growing grain might be the answer.

Representatives from five farms, a flour mill and several non-profit groups joined together to form the Heritage Grains Collaborative. They hoped that reintroducing heritage grains to the area would increase the region's food diversity and food security, while also providing for new economic opportunities.

Award-winning author and University of Arizona professor Gary Nabhan led the Collaborative. An advocate for sustainable agriculture and heirloom seed saving, Nabhan has been dubbed by *Mother Earth News* as "the father of the local food movement."

It was because of Nabhan's books, presentations and food awareness projects, such as Sabores Sin Fronteras (Flavors Without Borders), that farmers and food processors originally became interested in Arizona-grown grains.

In 2011, hopeful for change, Nabhan and the rest of the Collaborative set out on a two-year project to revive the production, milling, distribution and marketing of heritage grains.

Fig. 9.1: *No good project can get off the ground without a little round-table discussion and planning session. The Heritage Grains Collaborative is no exception.*
(CREDIT: NATIVE SEEDS/SEARCH)

Fig. 9.2: *Chapalote flint corn has a long history in Arizona. Generally brown in color, it can actually range quite a bit in shade.* (CREDIT: NATIVE SEEDS/SEARCH)

But if you are going to grow heritage grains that are no longer in mainstream production, where do you get the seeds?

Luckily, seeds from some heritage grain varieties were still available, thanks to a seed bank maintained by Native Seeds/SEARCH — which Nabhan had co-founded in 1983.

Native Seeds/SEARCH is a Tucson-based non-profit organization that preserves and promotes the use of the region's ancient crops, many of which are rare or endangered. Ultimately chosen for the project were White Sonora soft wheat and Chapalote flint corn.

The two grains are uniquely suited for the arid southwestern climate, as they often need less water to grow than modern grain varieties. What's more, they can help tackle the state's hunger issues because they are able to produce consistent yields without chemical fertilizers. Both grains have a small carbon footprint and are well-suited to a sustainable agriculture model.

The two grain varieties also have long histories in the region and are possibly the oldest varieties of their species to be farmed in the Arizona desert. "[Chapalote] is the corn that was grown here 4,100 years ago in the same valley," Nabhan says.

A flint corn similar to popcorn, Chapalote's deep brown color is unique. Traditionally, it was toasted, ground into flour and mixed with water to create an energy drink. More recently, however, it has been ground into flour to make tortillas.

As for the White Sonora, it's not as old, but it has an equally impressive history in the region. Spanish missionaries brought White Sonora wheat to the region in the late 17th century. The indigenous people at the time were dependent on the two cycles of corn that could be grown in a single summer and the harvest of native mesquite pods. They welcomed the introduction of a food crop that could be grown in the winter.

White Sonora was grown for generations and became America's first export crop. It fed both sides in the Civil War. However, when modern wheats became available, it could not compete with respect to productivity. While its yield was more predictable, crops yielded only about 30 to 50 percent of what more modern wheats could produce.

By 1975, neither White Sonora wheat nor Chapalote corn were commercially available in Arizona.

The Heritage Grains Collaborative has set out to change that.

With seeds in hand, Nabhan submitted a grant proposal to the Western Sustainable Agriculture Research and Education (SARE) program in late 2011. In March 2012, the Heritage Grains Collaborative received $50,000 in funding for a two-year research project.

In 2012, the Collaborative grew both grains and documented growing conditions such as the soil type and fertility, irrigation and yield.

Their minimal need for irrigation is what makes White Sonora wheat and Chapalote flint corn so well-suited for desert production, especially with the challenges of climate change.

According to Nabhan, three thousand counties in the US and five provinces in Canada had disaster areas in 2012, with devastating crop failures. "We're going to need more of these drought-tolerant crops," he says. "GMOs are not going to be bred quickly enough and be cheap enough to really help many people. We'd rather get proven drought tolerators in farmers' fields are quickly as possible."

Two of the growing sites in the region (Avalon Organic Gardens and Tubac Presidio State Historic Park) are thought to be where Jesuit Father Eusebio Francisco Kino planted Arizona's first crop of White Sonora wheat. Two farms belonging to another Heritage Grains Collaborative member, the Community Food Bank, are also believed to be within a few miles of where the country's oldest Chapalote-like maize was found.

Fig. 9.3: *A selection of the brownest Chapalote from the summer of 2012 was used as seed for farmer Kyle Young's 2013 crop.* (CREDIT: KYLE YOUNG)

It is stories like these that have inspired the Collaborative to see grain-related tourism as another potential solution to the region's economic challenges.

The region is moving towards being designated a US National Heritage Area, similar to some of the wine trails across the country that promote heritage and tourism. The more that people hear these stories and visit the area, the more economic spinoffs can be realized and jobs created.

To foster the economic aspects of the program, the Collaborative's public outreach, education and marketing efforts began in the spring of 2012.

As part of that campaign, members of the Collaborative spoke about the importance of local heritage grains and provided samples of foods prepared using the grains (such as bread and tortillas). Like many projects to bring back heritage grains and save them from extinction, such as the Ark of Taste, the Collaborative knew that having people taste these grains and incorporate them back into the mainstream food culture would be key to their revival.

To help educate potential growers and other stakeholders about the two grains, the Collaborative hosted grain schools: week-long introductory courses in heritage grains for small-scale farmers, bakers, chefs, agronomists and anyone else interested.

At the same time, the group also prepared a short booklet about grains and plans to create a larger book at the two-year project's conclusion.

Taking Things Slowly

Some North American wheats, including White Sonora, have earned inclusion on the Ark of Taste, a prestigious list of heritage foods maintained by the Slow Food movement. These grains qualify for the list due to their unique taste, their history in a specific region and their risk of becoming extinct. Slow Food advocates believe that these unique varieties will be revived if we grow, eat and come to see the value in these foods. North American wheats that have made the list include Red Fife and Turkey Red, as well as White Sonora.

In June 2012, the Collaborative celebrated its first harvest of 40 acres of grain.

While growing historical grains to combat climate change is essential, Nabhan says, addressing poverty has proven to be a more daunting task. For example, one of the project's original goals was to ensure an affordable nutritious food source.

"Grain usually doesn't cost as much to purchase as fish or meat or dairy products," Nabhan explains. The Collaborative believes that having an abundant supply of low-cost, nutritionally rich grains available in the region will make Arizona more food secure.

What's more, many people with gluten intolerances are able to consume older grain varieties like White Sonora wheat, says Tarenta Baldeschi of Avalon Organic Gardens and EcoVillage.

One of the Heritage Grains Collaborative's growers, the EcoVillage regularly provides 100 people with bread baked from heritage wheat. They've noticed something interesting.

Fig. 9.4: *Members of the Heritage Grains Collaborative survey the fields early on in the project's first year.* (Credit: Hayden Flour Mill)

Fig. 9.5: *Avalon Organic Gardens and EcoVillage regularly bakes bread for more than 100 people and reports their heritage wheat loaves yield few gastro-intestinal complaints.* (CREDIT: AVALON ORGANIC GARDENS AND ECOVILLAGE)

"Individuals who are allergic to wheat or gluten are able to consume this heritage wheat because its genetic composition is more true to nature than many new hybridized high-producing wheats," Baldeschi says.

Indeed, so far no allergic reactions have been reported.

"And with less money needed to purchase grains or bread, or to find alternative special needs foods, we are reducing poverty in this particular way," he adds.

While the Collaborative still has the goal of providing low-cost nutritious grains, "we're thinking that the link to poverty may end up being more in livelihoods," Nabhan says.

The Collaborative envisions local tortilla makers and bakers using the grains to launch new businesses while also feeding their communities.

"[Sonora] wheat is excellent for the large oversized wheat tortilla, which was initially developed in this part of the US/Mexico borderlands," Nabhan says.

These large tortillas are called *sobaquera,* which means "armpit tortillas": when being made, they stretch from fingertips to armpit. They're proving to be popular on the food van circuit.

"More and more people are just buying lunches and even dinners from mobile food wagons," says Nabhan. "They use a lot of tortillas, so we're trying to get the connection to get these [heritage grains] back into the fray."

But to make tortillas you first need a mill to process the whole grains into flour.

"We used to have about 60 mills within a 200-mile stretch on either side of the border in Arizona and [the Mexican state of] Sonora," says Nabhan. "We want to see some of those mills renovated, as is happening in other regions."

To support this revival, Nabhan is trying to identify borderland communities where there's a chance of renovating historic mills to provide jobs. "In some of the border cities, we have 25 percent unemployment right now," he says.

Mills will be a crucial part of the final picture.

Another key to the project's success will be having a strong value chain from producers right through to consumers, says Nabhan.

"The thing that's replicable is having a vertically integrated co-op where we have millers, bakers, chefs, food historians and farmers collaboratively planning things," he says. "That means that we're going to get out of sync less and less in terms of setting reasonable prices that really honor people's work at each stage in the group supply chain."

He says this integration of stakeholders is the best way to ensure that the grains get into the hands and mouths of the poor, as well as into the hands of processors and other local small-scale entrepreneurs.

In the meantime, the group is distributing grain seed to growers on

Fig. 9.6: *An abandoned flour mill in the agricultural village of Huepac in Sonora, Mexico.* (Credit: Native Seeds/SEARCH)

both sides of the US-Mexico border to encourage the growth of White Sonora wheat and Chapalote corn.

While the ebb and flow of the economy will ultimately drive its growth, Nabhan and the other Collaborative members believe the project is worth promoting because of the possibilities it brings in both jobs and food security.

Like many of the other community grain-growing projects across the continent, early success has been tangible despite a few hiccups along the way. After all, the Collaborative's members have successfully grown grain, and found ways to get it into the hands and mouths of some local residents. It's still too soon to tell, however, whether Arizona's Heritage Grains Collaborative model is one worth replicating on a larger scale or in other communities.

"We are just a year or so into the project, and we don't want to over-tout it," Nabhan cautions. "It's a good model, but we don't have enough results yet to say it's a successful model. It's just way, way too early."

Join the Revolution

- Research the history of grains in your region. Did local farms once grow Red Fife, Acadia, Red Turkey or another heirloom variety? If you can still find seeds for these varieties, try growing them to share with a seed bank or other growers.
- The Ark of Taste project believes that eating heirloom varieties is the secret to reviving them in mainstream food production, so make tortillas or bake bread with your heirloom grains and share with friends.
- Meet with other local grain growers after the fall harvest to compare experiences and best practices, connect with machinery and seed resources and maybe organize bulk purchases of seed.

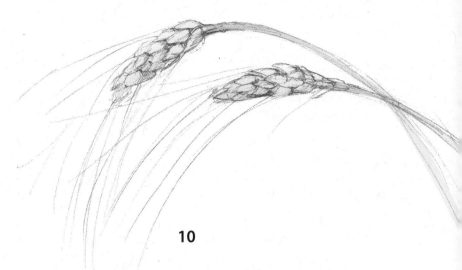

10

The Mendocino Grain Project: Grapes and Grain — California's Strange Bedfellows

No two substances in nature have to suffer more to become what they are than bread and wine.

— Fulton J. Sheen

NEBRASKA-BORN FARMERS JOHN GRAMKE AND DOUG MOSEL have joined with Sophia Bates of Apple Farm to reintroduce grain to Mendocino County, which sits at the north end of California wine country, about midway between San Francisco and the Oregon-California border.

But unlike the vast Nebraska grain crops of their childhood, the Mendocino Grain Project is growing some of its grain in significantly tighter quarters. With much of the region's farmland being used as vineyards, the Mendocino Grain Project is experimenting with growing grain between rows of grapes.

"We're growing grain in vineyards on a small scale, still testing it," explains Mosel. "We've done it with varying degrees of success three seasons in a row."

> "We're growing grain in vineyards on a small scale, still testing it. We've done it with varying degrees of success three seasons in a row."
>
> — Doug Mosel, farmer, Mendocino Grain Project

In 2012, year three of their project, the farmers planted over 60 acres of open-field grain and lentils, plus just under five acres of crops in ten acres of vineyards.

It seems like an odd practice, but Mosel and his fellow revolutionaries aren't alone.

"I know there are a handful of others who have grown in vineyards previously, if not contemporaneously with us," he says. "There are at least three vineyards that I know of who are now talking seriously about introducing this practice."

In fact, a nearby family grows most of their wheat exclusively in their vineyard. The grain is essentially a bonus crop, allowing the farmer to get more use from one plot of land.

"The land between the grapes is not being used anyway, so it makes for very good double use of the vineyard," Mosel explains.

Fig. 10.1: *Planting in the vineyard in the spring of 2009 meant smaller machines and tighter quarters.*
(Credit: Doug Mosel)

What's So Great About Cover Crops?

Cover crops lead to healthier soil. Healthy soil leads to healthy plants. Healthy plants can better fend off pests, they have stronger immune systems and so can recover from pest attacks, and they can attract other beneficial insects that attack the unwanted insects. As a result, pesticides aren't needed, because the plants can fend for themselves.

Acting as cover crops, grains help to prevent soil erosion, deter weeds and hold moisture in the soil. Grains can even reduce the use of pesticides. Mosel says some grains can resist some of the nematodes that can be a problem in vineyards.

And this symbiosis continues underground as the grains don't seem to harm the grapevine roots. The grains are planted in every other row between the grapevines.

"They don't interfere with the care of the vineyard, and because grains are not very thirsty, there's no issue of sucking moisture away from the grapes," Mosel says. "Because the grain matures well ahead of the grapes, the harvest is finished before there's a need to enter the vineyard to harvest the grapes." The farmers are careful to use harvesting equipment that fits between the grapevines to avoid damaging the plants.

Thanks in part to media coverage and a Facebook page, interest in the project has been substantial. The group encourages other farmers in neighboring counties to grow grain on plots from anywhere from a few dozen acres to upwards of a hundred.

At the outset of the project, the stated goal was to reintroduce local grain production to the county.

"Historically, grain was grown all over the region. There were even grains such as barley and oats grown on the coast here," Mosel says. "Inland, wheat, barley, oats and rye ... they were all grown until the mid-1960s to such an extent that they were recorded as agricultural products of the county."

Fig. 10.2: *The Wintersteiger threshing machine makes short work of the vineyard harvest.*
(Credit: Doug Mosel)

As in much of North America, the availability of cheap industrial grain, especially wheat, made it impractical to continue to grow grain in Mendocino County. A little place like Mendocino simply can't match the scale of growing in places like Montana and Saskatchewan. As a result, wine grapes became the crop of choice.

But grain is making a comeback. In the three years since the Mendocino Grain Project launched, community support has been remarkably strong.

"If we expand it from what we are doing in our county to the neighboring counties, regionally, it's phenomenal," says Mosel, who is well aware that the grain revolution is not limited to California.

"This is happening all over the North American continent — in various provinces of Canada, and in surprising places in the US where I wouldn't have expected it," he says. "There are projects all over."

He credits the local food movement for reviving interest and awareness. While the focus has been so strongly on produce, and more recently on meats, Mosel says there's a growing recognition of the absence of staple grains and the fact they play an important role in food security as well.

What's more, bakers are discovering the pleasure of baking with truly whole grain flours and looking for sources of local grains. In turn, "farmers have started to rise to that invitation to grow for them," Mosel says.

Parallel to this renaissance of locally grown grain has been the return to growing grains for sale in the local market, as opposed to growing for sale into the commodity market.

Because the demand for local grain is increasing in the region, the project's three main farmers will be contracting with other local farmers in the future to grow some additional crops.

Fig. 10.3: *Local grains are making a comeback in California wine country thanks to initiatives like the Mendocino Grain Project.* (Credit: Doug Mosel)

Mosel calls the project a "friendly collective of small-scale growers who are growing and distributing independently." The farmers share their experience and some resources, such as equipment and seed.

Moving forward, the Mendocino Grain Project hopes to follow the model of Oakland-based Community Grains, a chef-led private enterprise.

"They have two or three farms growing grains for them, and they're beginning to develop and market value-added products such as pasta," Mosel explains. "That's one of our eventual goals as well. I'm quite sure the way things are going all of that can and will happen here in short order."

Join the Revolution

- Go rogue and plant some grain seeds in an unusual space that could use livening up. Think plant pots, a neglected lot, a soil-filled divider between roads; grains need minimal watering and are ideal plants for guerrilla gardening.
- Growing wine grapes and grains together is one thing, but there are other ways to combine grains and alcohol. Is there a micro-brewery in your region that uses locally grown wheat, barley, rye or other grains? Raise a glass and celebrate local grain production.

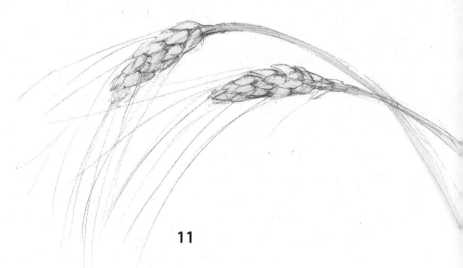

11

The Ripple Effect:
A Rock of Any Size Makes Waves

I F THERE'S ONE THEME ALL OF THESE TALES FROM THE FRONT LINES
have in common, it is the impact one single person or idea can have on
an entire community. Every small uprising not only creates its own ripples,
it's also part of the larger unsettling of what has become the status quo.

The community grain revolution is being fought on many different
fronts all over North America. Each individual battle helps shape the
larger changes.

The revolution is taking place in the fields, where some groups are
slowly growing out ancient grain varieties to help bring them back into
mainstream production. For example, the Heritage Wheat Project, which
connects research projects across Canada, was instrumental in the revival of
Red Fife wheat. The Heritage Grain Conservancy, based in Massachusetts,
has a special interest in emmer and einkorn.

Other revolutionaries are focused on trialing different grain varieties
to determine their suitability for different climates, such as the Backyard
Beans and Grains Project in Western Washington's Whatcom County and
the Northern New England Local Bread Wheat Project.

Meanwhile, eaters are beginning to rebel against the bland industrial bread at the grocery store. You can start your own uprising simply by switching to artisan bread from a local baker, or learning to make your own loaves. Then take a look at the ideas for a DIY uprising on the following pages to see if there's an opportunity for change in your community.

You may be able to join a grain or bread CSA, if there's one in your area. If one isn't already available, you can start one.

Anyone can start a CSA program. The owners of Wheatberry Bakery in Amherst, Massachusetts, began the Pioneer Valley Heritage Grain CSA. CSA programs have also been started by eaters, such as Vancouver's Urban Grains and, on the other side of the country, Ontario's Haliburton Grain CSA. Some grain farmers choose to sell their grains through the CSA model, including Country Thyme Farm in Bowden, Alberta, and the Prairie Heritage Farm in Conrad, Montana. The latter offers both bread and grain CSAs.

Some grain CSAs, such as that run by Lonesome Whistle Farm in Oregon's Willamette Valley, include lentils, dried beans and other legumes in addition to grains. This makes a lot of sense. Once a farm has the infrastructure to grow, harvest, thresh and store grains, much of that equipment can also be used to grow, harvest, thresh and store dried legumes.

Now that you've started down the revolutionary road, what next? It's time to grow some grain! Even if you don't plan to eat it, or if you don't have a spare square foot of soil, growing grain is possible and rewarding.

The Perks of Peas and Beans

Legume plants help increase the nitrogen in gardens and farmers' fields, and so are often planted as part of a crop rotation. Legumes are "nitrogen fixers," which means that they help convert nitrogen in the atmosphere into a usable form for the plant to use, and when the legume plant dies, that nitrogen is released into the soil. Planting legumes actually improves the soil, rather than just taking nutrients out of the soil as most other crops do.

Do you know what a wheat seedling looks like? A quinoa seedling? It's time to learn. And it's easy.

Grab a friend and find yourself a patch or pot of soil. Go guerilla gardener if you have to and suss out a corner of neglected land. Then get your seeds: order from a seed company, ask your local miller or just buy a bag of wheat kernels from the grocery store. If it's March/April or September/October, plant. (If it's not, wait awhile or plant indoors.) Make sure your seeds are watered or get rained on right away, then leave them alone and wait. Watch your grain grow.

If you have questions, the rest of this book gives you the information you need to choose, grow, tend, harvest, thresh, winnow and eat your own grains.

Give some extra seeds to a friend or two. Compare notes.

Welcome to the revolution.

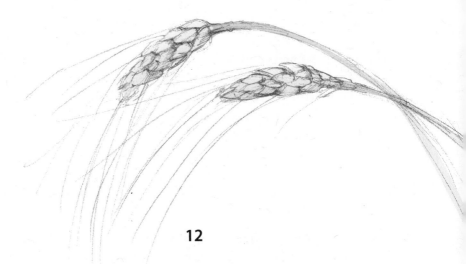

12

Join the Revolution:
Your DIY Uprising Checklist

You've now read the stories of communities across the continent whose residents decided to take control of their daily bread. Each mini-revolution has had its own ups and downs, but lessons can be learned from all of them. Here's a handy compilation of ideas you can use to create your own uprising, inspired by those "Tales from the Front Lines":

Vote with Your Fork

- Sign up for or start your own "eat local" challenge. Try to eat local food for at least one day each week for a month.
- Join a grain CSA. If there isn't one in your community, set up a meeting to gauge public interest in starting one.
- Flex your consumer muscles: shop with intention. Request that local stores carry local products and don't settle for non-local substitutes.

Knowledge is Power

- Educate yourself. Plant some grain seeds in your backyard for the simple purpose of seeing how they grow.

- Host a potluck and challenge your guests to bring a dish that has grain as an ingredient. Bonus points for those who find locally grown grain.
- Organize a farm tour and get to know your area's growers.
- Dig a little into the history of grain in your community. Was it being grown at one time? Is it still?
- Attend a Seedy Saturday event. If there isn't one near you, start your own.

Community Sufficiency

- Attend the Kneading Conference, or Kneading Conference West.
- Organize a local mini-conference or community artisan bread fair of your own. Invite local farmers, millers, bakers, brewers and eaters to connect, collaborate and, best of all, to break bread together.
- Buy bread that's made with locally grown and/or milled wheat and celebrate the tasty inconsistency.
- Rip up your lawn and plant grain. At the very least, it will get your neighbors talking.

Making Connections

- Research how you would feed yourself and your family if your traditional food supply was cut off. Get prepared.
- Find out where your nearest flour mill is. Ask where they get their grain from.
- Increase community sufficiency in your area by buying local products.
- Rally your neighbours or community and build a community oven. Organize a public planning meeting to connect everyone interested in building the oven, and invite a guest speaker who can talk about how other communities have succeeded. Take the names and contact details of anyone interested in being involved: bakers who can teach workshops, construction crew workers, marketing gurus and more. Every subdivision would become a neighborhood if residents gathered for regular pizza parties.
- Ask your local library to stock books about wood-fired ovens and bread baking. Even better: buy them and donate them to your library.

- Join a bread CSA. If there isn't one, considering starting a bread club (see Chapter 17).
- Put out the call to those you know who make bread and get a sourdough starter. Sourdough starters often have amazing stories from being shared and passed down through the generations. Or, follow the steps in the "Eating Grains" section to make your own sourdough starter. Then learn to make sourdough bread.

Thinking Outside the Box

- Research the history of grains in your region. Did local farms once grow Red Fife, Acadia, Red Turkey or another heirloom variety? If you can still find seeds for these varieties, try growing them to share with a seed bank or other growers.
- The Ark of Taste project believes that eating heirloom varieties is the secret to reviving them in mainstream food production, so make tortillas or bake bread with your heirloom grains and share with friends.
- Meet with other local grain growers after the fall harvest to compare experiences and best practices, connect with machinery and seed resources and maybe organize bulk purchases of seed.
- Go rogue and plant some grain seeds in an unusual space that could use livening up. Think plant pots, a neglected lot, a soil-filled divider between roads; grains need minimal watering and are ideal plants for guerrilla gardening.
- Growing wine grapes and grains together are one thing, but there are other ways to combine grains and alcohol. Is there a micro-brewery in your region that uses locally grown wheat, barley, rye or other grains? Raise a glass and celebrate local grain production.

Section Two

A Hands-On Guide

Get Your Hands Dirty

Do you or I or anyone know how oats and beans and barley grow?

— Raffi

RAFFI HIT THE NAIL ON THE HEAD with his popular children's song, "Oats and Beans and Barley." Most of us don't know how these basic foods are grown, which is why we've included this hands-on guide to growing your own oats, barley and other grains.

This guide is a collection of tips, technical details and nifty facts that we've gleaned from growing grains on a small scale since 2008, hosting and taking notes at numerous Island Grains workshops and reading a vast array of books and online resources. Given the extraordinary range of climate zones across North America, it's challenging to suggest planting or harvest dates, much less grain varieties, that will work for everyone.

Please keep in mind that we are not professional grain growers; it's quite possible that your wheat farmer uncle will refute something we've included here. That said, we hope this guide gives you some useful information to start with. The revival of small-scale grain production continues to evolve as more of us experiment in our own gardens.

The following pages may provide more details than you will ever want to know about how to grow grain on a small scale. While you're welcome to read this section all at once, it may be easier to use it as a reference as you progress through the various stages of growing grains.

Indeed, the best way to learn is often by doing it yourself. We urge you to find a square foot or more of soil, get a tablespoon of grain seeds, and start growing grain.

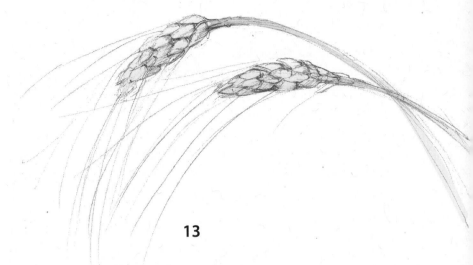

13

Grains 101: The Basics

*If you truly understood a single grain of wheat,
you would die of wonder.*

— Martin Luther

T**HE FIRST STEP IN GROWING GRAINS** is to get up close and personal with the grain plant itself.

Wheat, rye and other grains are all members of the grass family. Thousands of years ago, our ancestors chose the grass plants with the largest seeds and, instead of eating these seeds, planted them to grow the next season's crop. After many, many seasons, this selective breeding process resulted in the grain plants we see today.

A grain seed, or kernel, is made up of the bran layers (15 percent of the seed but 80 percent of its nutrients), the germ (5 percent of the seed) and the endosperm, which contains most of the seed's carbohydrates and proteins. The smallest part of the kernel, the germ, contains most of the antioxidants, vitamin E and essential oils. Any chemical inputs used to grow the grain, such as herbicides and pesticides, become concentrated in the germ.

If a grain seed is exposed to the right amount of moisture and warmth, the seed germinates. The plant's embryo (the germ) feeds off the nutrients stored in the rest of the seed (the endosperm) and starts to grow, first sending a little root through the protective bran layers.

Surrounding the grain kernel is a thin layer called the hull, or husk. Just as nuts have protective inedible shells, grain seeds have hulls. Like a nut's shell, the hull must be removed to make the seed kernel palatable and digestible for humans.

Fig. 13.1: *Awnless grains.* (Credit: Native Seeds/SEARCH)

Some grains, such as rye and most wheats, can be easily separated from their hull. This separation process is called "threshing," and it can be as simple as beating the grains with a baseball bat. Other grains, including emmer and spelt, require mechanical processing to remove the hull.

Some grain plants have "awns" or "beards," the barbed, pointy spears that grow from the seed head. These awns deter deer and other grain-loving animals from eating the seeds, and can make it more challenging for humans to handle the grains when harvesting. The barbed awns can prick and work their way into your skin, so caution (and a good pair of gloves) is recommended when handling bearded grains.

Each seed head grows on a stalk. When these stalks are dried they become straw, which is useful for feeding livestock, thatching a roof, mulching your garden or using as warm bedding for animals. Straw is hollow inside, which allows it to hold in heat.

Fig. 13.2: *Triticale grain, a hybrid of wheat and rye, with awns.*
(Credit: Hungry Ghost Bread)

Planning Your First Grain Garden

You could grow grains every season for the rest of your life and never have to plant the same variety twice.

There are tens of thousands of varieties of wheat alone, each with its own subtle flavor, soil and climate preferences, and pros and cons depending on your particular needs in the garden or kitchen. You could grow grains every season for the rest of your life and never have to plant the same variety twice.

One small seed of grain can produce hundreds of seeds in a single season, which means that you can start with a mere tablespoon of seeds and grow that tablespoon out to a pound, then tons, over a number of years. In this way, heritage varieties such as Red Fife wheat have been brought back from near extinction into mainstream production. This ability to exponentially grow out grain seed makes it possible to start with a few packets of purchased seeds and be able to grow enough grain to feed yourself within a few seasons.

Growing grain and saving some for next year's seed allows you to encourage certain traits, as humans have done for centuries. You can choose which plants to save the seed from each year, and thereby encourage the traits you want (e.g., stalk height, disease resistance, kernel size). Selective seed saving can ultimately develop that original variety into one that reliably meets your particular needs. Some grains, notably those known as landraces, adapt to different growing conditions easily and, over the seasons, can be influenced by the soil, local climate and even your personal growing practices to develop a distinctive flavor. This flavor is called "terroir," a term more commonly used for grapes or wine.

When growing grains on a small scale for food, the most important factor is whether the grain kernel can be easily removed from its hull or not. Some hulls can be very difficult to remove without machinery. Most wheats, barley and rye kernels are easily separated from their hulls, but spelt, buckwheat and emmer are impossible to hull without machinery or time-consuming hand processing. So-called hull-less or naked varieties of oats are available, but are still not always easy to process.

You will also want to think about what grains you use in your kitchen. If you often buy wheat flour, quinoa and whole grain rye at the grocery

For the Love of Your Soil

Grains aren't just grown for food. All grains help your soil while they're growing, too.

A grain plant's roots can go deep into the soil, collecting excess nutrients (most notably nitrogen) and storing them for the next crop to use.

Roots also break up the soil when they grow, improving the soil structure and helping water find its way deeper into the soil through cracks.

As grains grow they send out "tillers" or side-shoots that grow into separate plants. In this way, a single plant can quickly occupy a square foot or so of growing space and make it harder for weeds and other plants to grow.

The tall grain stalks also block out sunlight, shading any weeds or competitor plants.

Finally, grain stalks soften the impact of rain and wind, while their strong roots hold on to the soil. In both these ways, grains prevent erosion and the loss of nutrients from your soil.

store, for example, it makes sense to grow these grains because you're more likely to eat them.

While many grains look similar, they each have different benefits and qualities that can help your garden and open the door to new experiments in your kitchen. To help you choose which grains to grow, here's the general lowdown on your options.

Cereal Grains

Wheat

Triticum vulgare
1 bushel = 60 pounds

Wheat has fed human beings for thousands of years — it really is the "staff of life." It's a household staple, useful for bread, pasta and pastries. The whole grains, called "berries," can also be cooked like rice, or even sprouted. (Check out the chapter on eating grains for more information.)

Wheat varieties are often described as "winter" or "spring" wheat, which indicates the time of year when that variety is meant to be planted. Winter wheat varieties, for example, are planted in the fall and harvested the next summer. Spring wheat varieties are planted around the time of the last spring frost, then harvested in late summer. Generally, winter-planted wheat can be harvested earlier and will result in a larger yield than spring-planted wheat.

The color of the seed's outer bran layer determines whether it's referred to as "red" or "white." Red wheats are said to have better, stronger flavor than white wheats.

Grains are further described as "hard" or "soft," which refers to the amount and types of gluten in the grain seed's endosperm. Hard wheats are higher in gluten; commercial bakers prefer them because the flour can take in more oxygen to create a lighter bread loaf. Hard wheats also hold more liquid than soft wheats, so less flour is needed when making bread. According to the Kootenay Grain CSA, 80 percent of all wheat grown in Western Canada is a variety of hard red spring wheat.

The hardest wheats are called durum wheats. They are very high in gluten and are used to make pasta.

Soft wheat flours produce denser bread than what is typically produced by the mainstream bread industry, but are preferred for pastries, cookies and other non-yeast baking, such as crackers and flatbreads.

Growing Wheat

Wheat grows best in dry, heavy loam or clay soil that isn't too fertile. Otherwise, the plants may grow too tall and risk falling over, or "lodging," which makes harvesting more difficult and can compromise the grain kernels' quality. A cool, moist spring and dry, hot summer are ideal.

About $\frac{1}{40}$ of an acre (approximately 1,100 square feet) of sown wheat can be expected to yield roughly 60 pounds at harvest time. It's easy to grow a few pounds of wheat on a garden scale, and the kernels can be separated ("threshed") from the less palatable hulls and plant matter ("chaff") using objects found around the home, such as pillowcases, baseball bats and shoes. Grain kernel size varies depending on the variety: for example, khorasan grains are noticeably larger than other wheat grains. The height of the stalk also varies, so if you want more straw consider growing a long-stalked variety such as Red Fife.

Khorasan

Triticum turanicum

Khorasan is sometimes referred to as Kamut, which is a trademarked brand name for organically grown khorasan. It is an ancient wheat variety, and is rumored to have been found in King Tut's tomb. (The brand name Kamut is from an old Egyptian word for wheat.) Khorasan is very high in gluten. According to the Speerville Flour Mill, khorasan is actually an old variety of durum wheat. It makes great pasta, and can also be used for bread and pastries.

Spelt

Triticum spelta

Spelt is one of the oldest cultivated grains and is higher in protein than many other wheats. It is quite similar to modern hard red wheat, but is said to be easier to digest. That may be because it is such an old variety, and/or because it is lower in gluten than many other wheats. It also needs less kneading than other wheats because its gluten is so fragile: bread machines can easily over-knead spelt dough. To avoid over-kneading, gently work the dough until it's elastic and smooth. Spelt flour is wonderful for sourdough baking.

Spelt sounds like a miracle wheat for anyone with gluten sensitivities, but be warned before you attempt to grow it on a small scale. It is very difficult to thoroughly remove the hull from spelt grains without machinery. You may end up with inedible, teeth-cracking, bitter-tasting grains. As well, the mechanical threshing process can damage the grain kernel and make it difficult to sprout, which limits its uses in the kitchen.

Anonymous Wheats

Many wheats, especially the modern varieties, are anonymous — instead of having a proper name like Turkey Red or White Sonora, they are simply described. For example, you can grow "soft white spring wheat" or "hard red winter wheat."

A notable difference between modern wheats and the older varieties is consistency. Because modern wheats have been bred to be harvested with the combine, they are usually a consistent shorter height. For example, a field of modern wheat plants may stand at an even two-foot height, while a field of Red Fife wheat has stalks from three to five feet tall. Consistent heights make it easier to harvest by hand too.

Landrace Wheats

Landraces are wheats that, while they may fall into a general family such as "Red Fife wheat," can still adapt to different growing conditions and develop some qualities distinct to the place where they're grown ("terroir"). For example, after a few years of planting, Red Fife wheat grown on a West Coast farm may have a subtly different flavor or appearance than a Red Fife wheat grown on an East Coast farm, even if both crops originated from a single batch of seed harvested on a Saskatchewan farm.

Other Cereal Grains

Rye

Secale cereale

1 bushel = 50–60 pounds

Just like rye bread, whole rye grains have a very distinctive flavor. Because rye is low in gluten, it is often used in combination with wheat flour when baking to make a lighter loaf. Rye can also be used to make whiskey. One versatile recipe to try is rye berry salad: the whole rye kernels are cooked like rice, then cooled and mixed with olives, sautéed onions and garlic, chopped vegetables, feta cheese and salad dressing.

Rye can be planted in the fall throughout most of North America, around the time of the first frost. It can survive dips in temperature to as low as –40 degrees Celsius, and can even survive under snow. Although its official per-acre yield is less than that of wheat, rye can produce crops on poorer soil than wheat and tolerates cold, drought and dampness better.

In addition to its many food uses, rye is often grown to improve the soil. It offers the usual benefits of preventing soil erosion and suppressing weeds, and rye is also the best grain for collecting excess nitrogen from the soil and

Fig. 13.3: *At six-foot-four, farmer Brock McLeod is a good measuring stick for this healthy rye patch.* (CREDIT: HEATHER MCLEOD)

storing it on its roots through the winter. As a result, whatever crop you plant next will benefit from having that nitrogen readily available in the soil. Rye is allelopathic — as it grows, the plants produce a natural biochemical that prevents the nearby growth of dandelions and Canada thistle.

Because it's often planted in poor soils in order to improve them, reported rye yields vary significantly. Nonetheless, it's reasonable to expect a harvest of 60 pounds per 1,100 square feet, the same as wheat. Rye stalks can grow up to eight feet tall (and yield a lot of straw), producing eight-inch-long seed heads with abundant kernels. Because of its height, there is the risk of the grain lodging in rainy or windy weather. Rye is very, very easy to hand-thresh.

Barley

Hordeum vulgare
1 bushel = 50 pounds
Barley doesn't have enough gluten to make bread unless mixed with other flours, but it's excellent when added to soups and can be used to make beer. "Pearl" or "pearled" barley has had the bran layers removed.

Barley seed heads have six or two rows of kernels, depending on the variety. Six-row varieties have more kernels per plant.

"Bearded" varieties may be deer-resistant once the stalks produce seed heads, since the long pointy awns itch and can stick in an animal's throat. Beardless varieties are also available and more pleasant to thresh by hand.

Barley excels with a long cool ripening season and moderate moisture but adapts well to hot, dry weather. It also tolerates salty and alkaline soils better than most grains. Spring-planted barley ripens faster than wheat. A harvest of 30 pounds of oats from 1,100 square feet is reasonable.

Oats

Avena sativa
1 bushel = 30–35 pounds
Oats beat out all other cereal grains when it comes to high protein and fat content. They can be rolled, ground into flour, baked whole into cookies or muffins and even soaked to create oat "milk."

Fig. 13.4: *When growing, oats looks very different from other grains.*
(Credit: Heather McLeod)

For spring oats, the earlier you can plant the better. As Gene Logsdon says, "Whenever the mud dries enough in spring to be workable, plant your oats." Aim to plant a week before the last spring frost. Oats like cool weather and don't need lots of sun. They require more water than other cereal crops to yield a good harvest, but have fewer insect enemies than corn or wheat.

Now that you're all excited about oats, here's the problem: most oats come with a very well-attached hull, and so are difficult to make edible. There are some clever hulling techniques out there, but your best bet is to plant a "hull-less" or "naked" oat variety. These varieties still have a hull, but it is much easier to remove. Before you plant your entire garden to oats, you may want to try out a few different "hull-less" varieties to make sure you can actually thresh and eat what you grow.

Whole oat grains can be fed to chickens, rabbits and pigs, who don't mind eating hulls, and oat straw is the best straw for feeding livestock —

it's even nutritionally preferable to poor-quality hay (dry grass), according to John Seymour's *A Guide to Self-Sufficiency*.

Pseudo-Grains

Neither amaranth nor quinoa are members of the grass family and so are not cereal grains, but they share some traits with grains in the garden and kitchen and so are considered pseudo-grains. While they require slightly different harvesting and threshing methods than true grains, both amaranth and quinoa can be grown on a small scale and are very tasty garden crops. The green plants can be eaten as well as the seeds. Note, however, that if you harvest too many leaves the plant may not be able to regrow.

Amaranth

Amaranthus

Amaranth was a sacred food of the ancient Aztecs, and no wonder: it's a complete protein and has more calcium per 100 grams than milk. The seeds are gluten-free, have a distinctive nutty taste and can be cooked to make a breakfast cereal. Amaranth seeds are very tiny. One pound contains an estimated 250,000 seeds. (One pound of wheat, on the other hand, contains an estimated 14,000 seeds.)

Amaranth excels in a very dry summer. If it rains during the final weeks before harvest, the seeds can sprout while still on the stalk and the harvest will be lost. In warmer regions, amaranth seeds can survive through the winter and sprout into new plants in the spring. Some consider it an invasive weed, and in fact, it's related to pigweed.

Sowing Seeds of Rebellion

If you grow quinoa or amaranth, consider yourself part of the grain revolution story. Growing and possessing these pseudo-grains was forbidden when the Spanish conquistadors arrived in Central and South America in the 1500s. Fields were destroyed, and anyone caught with these seeds was severely punished. Nonetheless, these seeds survived.

Quinoa

Chenopodium quinoa

Quinoa originates in the Andes and was sacred to the Incas, who called it "the mother of all grains" or "the mother grain." It is a complete protein, very high in iron and calcium, practically gluten-free and needs less water than rice to cook.

Fig. 13.5: *Quinoa in bloom on Canada's West Coast.* (Credit: Helen Reid)

Quinoa is related to lamb's quarters, a succulent and tasty weed. It can grow to be five feet tall and produces beautiful flowers, making it an attractive pseudo-grain for flower gardeners.

Even More Grains

There are many more grains you can experiment with, including millet, emmer, sorghum, rice and corn, as well as pseudo-grains such as buckwheat. We don't recommend growing emmer or buckwheat on a small scale because it's just too darn hard to remove the hulls and make your harvest edible. As for millet, sorghum, rice and corn, it might be wise to pick up a copy of Gene Logsdon's *Small-Scale Grain Raising* or one of the other great resources out there once you're ready to try your hand at these crops. (See the "References" section at the end of this book for some recommended reading.)

Suggestions

Try growing grains that you'll actually use, such as wheat to grind into flour and bake with. To get you started on discovering which grains are right for you, here are some recommendations.

Fig. 13.6: *Emmer is notoriously hard to thresh on a small scale.* (CREDIT: JESSE COTTINGHAM)

Gateway grains, which are easy to harvest and cook with:
- rye (try making rye berry salad using the recipe in the "Eating" section)
- any wheat except spelt (grind it into flour, or sprout the whole grains and make Essene bread)
- quinoa (use it instead of rice as a complete protein side dish)

Beautiful grains, for gardeners or bouquets:
- blue wheats, such as Utrecht Blue
- flax (the plants produce pretty blue flowers, then the seeds form inside little golden balls that jangle when dry and touched)
- quinoa (stunningly beautiful plants)

Practical grains for homesteads and farms:
- rye (a wonderful green manure and cover crop that builds soil fertility; the tall stalks produce lots of straw for bedding, mulch or thatch)
- oats (to feed your livestock)

Heritage varieties:
- Different grains used to be grown in different regions of the continent. Check with a local historian to see if a particular grain once thrived in your community. For example, Acadia wheat is a heritage grain in the Canadian Maritimes.

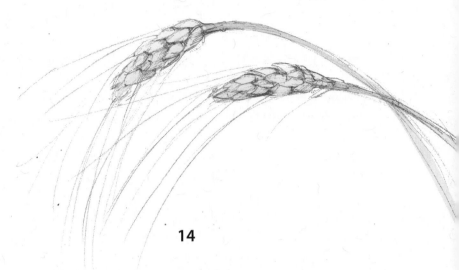

14

Growing Grains: Digging In

The first question that naturally comes to mind for gardeners consider-
ing grain growing is why one would think of growing wheat, barley
or oats given the millions of acres that big-time agriculture devotes
to grains in North America. My answer is that growing even a small
patch of grain can give you an intimacy and appreciation for some of
our basic foods that will last forever.

— Dan Jason

CONGRATULATIONS ON TAKING THE PLUNGE and deciding to grow your own grains. Now that you've decided which grains you'd like to grow, you probably have some questions. Here are some answers that will hopefully set you on the path to a great first harvest.

Where should I plant my grains?

Grains can grow too tall and risk falling over if grown in too-fertile soil, so if you've been feeding your garden with compost, let your vegetable crops have those areas and stake out the muddy, dusty or other marginal parts of your yard for your grains. Growing wheat, oats, rye and other cereals is an opportunity to improve the soil.

On the other hand, quinoa and amaranth aren't grasses and need a little more love, so consider giving these pseudo-grains some space in your normal garden.

While all grains grow best in full sun, they range across the spectrum when it comes to soil and climate preferences. Wheat likes heavy loam or clay soil. Oats do just fine in wet, cold soil. Rye grows well in dry, sandy soil, and does better in acidic soils than other grains. Barley likes warm, dry, lighter soil and grows just fine on even less-fertile soil then wheat.

This is a good time to mention that grains need little to no watering, and in fact they'll need to dry down completely by summer harvest time. If you interplant your grains with vegetables that need regular watering, one crop or the other will suffer.

Grains grow upwards of two feet tall, and may shade out other plants depending on where they're planted.

Where do I get seeds?

Established seed companies are a perfect source for reliable grain seeds. Many offer heirloom or unique seed varieties. The downside is that they usually sell small amounts and charge more per pound than if you buy seed in bulk, but you can easily grow a few tablespoons of grain seed out into multiple pounds within a few years. Search online for a seed company that sells grains suitable for your region.

In North America, we're lucky to have seed banks and seed-saving organizations that preserve unique, rare or heirloom seed varieties. These organizations may give you seed to plant in exchange for a share of your harvest so they can replenish and grow their supply. Some seed banks include Native Seeds/SEARCH and the Seed Savers Exchange in the US, and Seeds of Diversity and the Seed and Plant Sanctuary of Canada. The Heritage Grain Conservancy in the US specializes in heirloom grain varieties.

Your local grain farmer, miller or baker is another helpful local resource. They may be able to supply you with multiple pounds of grain seed, and if they don't know whether it's a fall or spring variety, or whether it's easy to hand-thresh, they can probably connect you with a farm or supplier who does have those answers.

When all else fails, head to your local grocery store. Some grocers, especially health food stores, offer whole wheat or rye berries, quinoa and other plantable grains. Any grain that can be sprouted is safe to plant, which means that processed grains such as rolled oats or flaked wheat will not grow in your garden. Spelt, oats and emmer grains may have been damaged when they were mechanically threshed and so may not germinate.

Keep in mind that, since grocery store grains are intended to be eaten rather than planted, important information on the grain variety isn't usually provided. For example, the emmer or wheat you buy may be a variety that's supposed to be planted in the fall; if you plant it in the spring, it may not grow as it should.

Another potential source of grain seed is your local feed or farm supply store. These grains may be suitable for animal feed, or intended for use as green manures or cover crops, and not as seed for human-grade food grain. The store may be able to answer questions about the quality of the grain seed, as well as whether it's suitable to be grown for food.

How much seed do I need?

Growing food is always an experiment, with numerous factors affecting the quality and quantity of your yield. How much grain you harvest at the end of the summer will depend on your soil fertility, climate, the weather throughout the growing season, how closely you space your seeds, how often you water (or it rains), weed pressures, where you source your seeds and many other influences.

If you just want the experience of growing grain, use whatever amount of seed you want. (Aim for at least ten seeds of each variety.) Any size of growing area — even a large pot of soil or a few square feet in your backyard — will make for a rewarding, interesting experience.

If your goal is to grow enough grain to supply your household needs, you'll need a larger area. For example, 1,100 square feet of growing space is about right to grow 60 pounds of wheat. To seed an area this size, you will need two to three pounds of grain seed (wheat, rye, barley and/or oats).

An 1,100-square-foot plot may sound large to some urban growers, but it's a realistic size for a single grower to manage and get a useful yield. Two people can harvest 1,100 square feet of wheat with pruning shears and scissors in a day. Threshing that grain into clean, usable kernels by hand takes even more time, but the amount grown on a plot this size (60 pounds of wheat) is manageable. An 1,100-square-foot plot is a realistic size for a first attempt at grain self-sufficiency.

If you broadcast your grain, which means tossing the seed out so that it falls haphazardly, you will need three or four times as much seed. You may also want to use more seed if you plant later in the fall or earlier in the spring, to ensure enough seeds germinate and to make up for seeds lost to hungry birds and other seed-eating creatures.

If you can, avoid planting all of your seeds, especially if you have a limited supply or are planting rare or unique varieties. If you keep a reserve supply and your crop fails, you'll be able to start again next season.

Is it better to plant in the fall or in the spring?

Winter wheats and fall rye are sown in early fall, grow a little and then go dormant during the cold months. When spring returns, they shoot up. This head start may result in a stronger plant and therefore a better yield (approximately 20 percent) at harvest time. You may even be able to harvest a few weeks earlier (e.g., July/August instead of August/September) than if you'd planted in the spring.

Spring wheats, oats and barley are usually sown in the spring. These grains are sown about the time of the last serious spring frost and are ready for harvest between June and September, depending on where you live in North America. When planting in the spring, sow your seed as soon as the ground can be worked: every week of delay will mean a smaller yield.

Some grains can be planted either in the fall or in the spring, while others usually won't survive cold winter temperatures and so must be spring planted (e.g., barley, quinoa, amaranth). If your winters are mild, you may want to experiment. Try planting barley in the fall, or plant half

Fig. 14.1: *You can sure tell the difference between the fall-planted hard white spring wheat on the left, grown from the same batch of seeds as the spring-planted wheat on the right.* (CREDIT: HEATHER MCLEOD)

of your favourite spring wheat variety in the fall and the other half in the spring so you can compare.

Note that while you may be able to get away with planting a spring variety in the fall, winter varieties planted in the spring may not mature in time to harvest, since these varieties may require a longer growing season. Winter wheats tend to grow best in temperate climates, while spring wheats like hot, dry growing conditions.

When should I plant grain?

Given how varied our climate regions are in North America, these are approximate times to plant. Experiment to find what works for you.

- Oats: as early as you can get onto the soil in the New Year. Possibly late February, depending on your micro-climate and soil.
- Spring-planted rye: March.
- Barley: March-May. Barley grows best in warm, dry soil.
- Quinoa: April. Quinoa doesn't like being wet.
- Spring wheats: February-April, or as soon as your soil can be worked. Spring wheats need a fairly long growing season.
- Flax: April.
- Khorasan: April.
- Amaranth: May.
- Winter wheats: September-October, or around the time of your first frost.
- Rye: September-October, or around the time of your first frost.

Do I need to do anything special to prepare the soil for planting grains?

Ideally, you should start preparing the soil six months or more before planting grains. Any sod should be removed or tilled into the top six inches of soil. After tilling with a tractor or rototiller, plant peas, beans, fall rye or another green manure crop to help break up the soil and improve fertility. This crop (and any grass or weeds) can then be tilled back into the soil when they are still green. Till once more immediately before planting your grain seed, to discourage any weeds.

Wheat and rye grow well in soil with small clumps, rather than powdery soil. The clumps will encourage rain to enter into the soil, rather than wash it away. However, barley likes a finer soil.

Wireworms, the larvae of click beetles, can be a major pest for grain growers — especially in newly tilled grass. These little orange creatures live the good life in the roots of grass plants, and will head straight for your grain seeds when they're planted. Wireworms will burrow into your seeds and devour them from the inside, making growth impossible.

If you're concerned about wireworms, plant into soil that's been tilled and grown in for several years, rather than freshly tilled grass. Or scrape the grass off the growing area and compost it separately, tilling the exposed soil for planting. But if wireworms decide they like your grain seed, there's not much you can do to stop them. Even non-organic pesticides don't work against these critters.

How do I keep track if I plant different grain varieties, or grain from different sources?

If you have seeds from different sources (e.g., farms, seed companies), keep them separate when planting — even if they are supposed to be the same variety. For example, if you plant Turkey Red wheat from two different seed companies separately, and the Turkey Red from one company germinates better, grows taller or yields more grain than the other company, then you know to order from that company again.

Fig. 14.2: *Planting in rows helps the grower tell the wheat seedlings apart from the grass. It also makes it easier to weed without stepping on tender young plants.*
(Credit: Heather McLeod)

As well, don't mix different generations of seeds when planting (e.g., seeds harvested this year vs. harvested last year). If one patch fails, this will make it easier to diagnose what went wrong.

How do I plant grain?

Planting in rows, by hand or with a row seeder, makes growing grain much easier. (You can also use a grain drill on a tractor, if you have access to one and are seeding a large area.) Rows use less seed and make it easier to track where you planted which varieties.

When growing grains, it can be tricky to tell your plants apart from the weeds. Since grain seedlings look just like grass, rows make it easier to tell the two apart so you can weed out the invaders. Similarly, quinoa is identical to its cousin, lamb's quarters.

If you have lots of seed and don't think weeds will be a challenge, you can broadcast it just like grass seed, sprinkling it out onto the soil in sweeping arcs. This is a much faster, easier way to plant, but you will use more seed this way and it can be difficult to identify or pull out weeds once the plants start growing.

Regardless of whether you opt to seed in rows or by broadcasting, you may want to leave room for a path so you can walk through your grain patch while it's growing and avoid treading on fragile seedlings. A path will make it much easier to get at weeds without damaging your plants.

A Grain Garden in Seven Steps

1. Till or pitchfork the soil.
2. Use a rake handle to draw horizontal rows, with enough space between the rows to fit your favourite hoe's blade.
3. Drop a seed every inch or two in the rows.
4. Cover the seed rows with a half-inch of soil.
5. Stamp the soil down with the rake's blunt head.
6. Turn a sprinkler on for two hours if rain isn't in the forecast.
7. Over the next few months, pull out weeds.

The war on weeds

Ideally, plant your grain seeds immediately after tilling the soil. Tilling will kill or at least deter weeds from growing, giving your grain seedlings a fighting chance as they compete for sunlight, water and the nutrients in the soil.

You may want to experiment with sprouting your grain seeds before planting, then tilling the seedbed and planting as soon as the kernels have sprouted. This will give your grains a solid head start on the weeds. (Refer to the "Eating Grains" chapter for sprouting directions.)

Spacing

Don't crowd your plants! Try to ensure even spacing of at least one to two inches between seeds. If broadcasting, toss the seeds out while walking quickly for the first pass, then adjust your pace for the second and third passes until the seeds are spaced properly. Finally, drop extra seeds in the bare patches. The plants will fill in the gaps with the tillers (side-shoots) that they send out, stifling weed growth and producing a greater yield at harvest time.

If you want to maximize tillers, plant your seeds in a row or staggered grid, with up to two feet between each seed. This spacing gives your plants room to spread out without interfering with each others' root systems. Encouraging your seedlings to tiller might delay the harvest, but it should increase your yield and you'll use less seed.

Depth

After planting, cover the seeds with a few inches of soil or rake the seeds into the soil bed. How deep you plant your seeds depends on your soil type and how much moisture from rain or irrigation the seeds will get those first few days. Seeds need moisture to germinate, whether it's already there in the soil bed or whether you turn on the sprinkler for a few hours after planting.

As a general rule, plant grain seeds half an inch deep in heavy clay soil or up to two inches deep in sandy soil. If you're unsure, err on the side of planting too close to the surface. If you plant the seeds too deep, they may not be able to access moisture in order to germinate, or may not have enough energy to grow up to the surface.

Tamping the soil

Tamp the soil down with the flat side of your rake or a roller to ensure good contact between the seeds and the soil.

Moisture

If the soil isn't already moist when you sow your seeds, plant when the forecast calls for rain or turn on the sprinkler to ensure your seeds get moisture to help them germinate. A week after planting, dig up a few grains to see if they're sprouting. (And to ensure they haven't been eaten by pests or critters.)

Should I keep records of some sort when I plant?

Yes. Label everything. If you're planting different varieties, mark those rows or sections if possible, but keep in mind that paper will deteriorate, while signs can fade and wander. When you plant, draw a map and note each variety's name, source and date of planting. If you know the history of the seed (e.g., who grew previous generations and where), then record that as well. Keep your map and notes somewhere dry and safe. You'll be glad you went to this extra effort many months later, when you harvest.

What can I plant with my grains?

According to the companion planting bible, Louise Riotte's *Carrots Love Tomatoes,* chamomile increases wheat yields (plant one part chamomile to every 100 parts wheat). Bachelor Buttons (a flower) is believed to aid rye production when planted in a 1:100 ratio.

Remember that grains grow two or more feet tall and require little water, so don't expect your companion plants to perform their best given this shaded, dry environment.

How much should I water my grain?

Grains are a wonderfully low-maintenance crop. Aside from some moisture to help the seeds germinate (which happens in a few days), grains don't need to be watered. If there's an especially hot, dry day, then feel free to provide some water, ideally through trickle tape or a drip irrigation system rather than a watering can or overhead sprinkler. Rye plants needs less water than wheat, which needs less than oats. Too much water may cause grains to grow too tall, at which point they might fall over.

Quinoa is traditionally grown in regions with less than three and a half inches of rain per year. Drought actually produces larger seed heads.

Once the seed heads form and the kernels mature, the grain plants need to die and dry down in order for the seeds to be dry enough to harvest and go into storage. Watering your grain plants at this stage will sabotage the process.

Will I need to do a lot of weeding in my grain patch?

It depends. If you planted in a new growing area that had weed seeds in it, things could get hairy. Planting in rows (rather than broadcasting) will help you tell the grain seedlings from weed invaders, and will make it easier for you to remove the weeds.

Once your grain plants are established, they can grow tall enough to shade out other plants, and so weeds should not be an issue at this point. If a weed does appear, remove it while it's still small.

Can grains be grown organically, or will I have to use chemical fertilizers, herbicides or pesticides?

Grains generally don't need fertilizers, and are often used as "green manures" themselves, which means that they're planted for the purpose of improving poor soils. In fact, if grains are planted in too-fertile soil, they may grow too tall and fall over.

Herbicides aren't necessary because grains are excellent at combating weeds all on their own.

While all grains can be grown without the use of chemical fertilizers, herbicides or pesticides, some grains (including spelt and khorasan) are said to be especially easy to grow organically. Even some conventional farmers who usually use chemical inputs don't bother when growing these grains.

Can I eat my grain plants?

Go ahead and make juice from the wheat grass, but do it around 14 days after germination and cut the grass to leave a few inches of the plant's stalk. At 14 days, your wheat seedlings will have lots of nutritional benefits to offer you, but will still be able to grow back.

It's best to press the grass into juice rather than shredding them in a blender, if you have an appropriate juicer.

The young leaves of amaranth and quinoa are both delicious and nutritious (both are rich in calcium and iron). When the leaves get older, you can steam them to make them more palatable. Don't over-harvest these plants: leave the stalk and enough leaves for the plant to keep growing.

How can I tell if my grains get a disease, and should I be concerned if they do?

Disease-infected grain plants may be stunted and not produce as much seed as healthy plants. Some plant diseases such as ergot can cause illness or death in the people or animals that eat them. Of all the cereal grains, wheat is the most susceptible to diseases.

Grain diseases are often spread as spores by the wind or water, or by planting infected seed. If your grain seeds show symptoms of disease, as a general rule don't use those kernels as seed for future crops. If your grains are afflicted with Fusarium or ergot, play it safe and don't use them as food for humans or animals either.

Rust

If your wheat or barley plants look rusty, your grains likely have rust. Rust is a fungal infection, and appears as red, brown, orange or yellow spots or streaks on the plant's stalks and leaves. If the rust spreads to the seed head and kernels, play it safe and destroy the grain.

Smut

Smut affects wheat and rye. The first signs usually appear once the seed head has formed: "smutted heads" stay green longer than unaffected heads and have a greyish tinge. The kernels themselves turn into "smut

Fig.14.3: *The discoloration on the wheat plants is rust.* (Credit: Brock McLeod)

balls." They are more rounded than healthy kernels, and will eventually release smelly black fungal spores.

Of the many varieties of smut, some are eaten by humans — one being corn smut, or "huitlacoche" as it's known in Mexico.

Dwarf bunt

Winter wheat is the most common victim of this fungal infection, but dwarf bunt can also affect spelt, rye and winter-sown barley. Like smut-infected grains, the plant will produce "bunt balls" instead of seeds. Bunt balls turn from green to brown, then crack open and release black spores. Sick plants may be shorter than healthy plants and may smell fishy.

Fusarium

Wheat, barley and corn are most commonly affected by Fusarium head blight, which also goes by the name of wheat scab, or tombstone disease, but it can also infect oats and rye. Watch for shriveled, chalky-white grain kernels, or grains with orange, black or dark purple marks.

Ergot

Ergot is most common in rye, but may affect wheat and barley, as well as a popular hybrid of wheat and rye called triticale. Watch for dark purple or black specks on the grain kernels when they're growing and maturing, especially if the weather is damp. Eventually the kernels will be replaced by hard black lumps. Never, ever eat ergot-infected grain.

Should I practice crop rotation with grains?

Crop rotation is a responsible practice if you want to keep your soil healthy and full of nutrients, no matter what you're growing. It means that you never plant the same crop in the same area in subsequent years, and that you vary what you plant to ensure your soil has the nutrients it needs. For example, you may plant wheat one year, then, once it's harvested, plant legumes such as clover, peas, beans, alfalfa or soybeans.

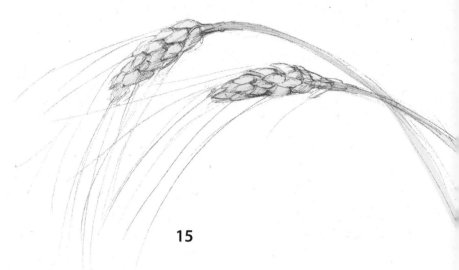

15

The Harvest: Reaping the Rewards

*If you wait until the wind and the weather are just right, you will never
plant anything and never harvest anything.*

— Ecclesiastes 11:4

When do I harvest my grain?

IF YOU HAVEN'T ALREADY LEARNED about the importance of the
weather while growing grain, harvesting season will bring that lesson
home. They say to "make hay while the sun shines," and this holds true for
grains as well. If your grain is too wet when it's harvested, it may not dry
down enough to store without risking mold or deterioration. Quinoa and
amaranth can even sprout while still on the stalk if the seed heads get wet
at harvest time.

So watch the weather forecasts and aim to harvest your grains after at
least a week of hot, sunny weather to help them dry.

Wheat, rye and barley

The goal is to harvest your grain seeds when they're ripe and dry, but not so dry that they "shatter," or fall off the stalk as soon as you jostle or touch the plant.

To test your grain kernels to see if they're ready to harvest, remove some grains from the seed head and try to dent a kernel with your fingernail. The seed should be hard enough that you can't make an indent. If you bite down on a kernel, it should crunch between your teeth.

If you can still make a slight impression in the kernel but you want to harvest it early (e.g., if there's rain in the forecast), harvest it but then let the seed heads dry out completely in a greenhouse or a dry, hot place until the kernels are hard.

Harvest times vary across North America and depend on when you planted, your climate and numerous other factors. That said, rough estimates are as follows:

- Barley: June to September, when the seed heads have bowed over and it's easy to remove the seeds from the plant by hand.
- Fall-sown wheat: July to early August. (Spring-planted wheat may be ready a few weeks earlier.)
- Fall-sown rye: July to August.

Oats

Harvest when half of the plants' leaves are still green, and you can dent the grain with your fingernail but not squish it. Check in on your oats and watch for them to mature from July through September.

Flax, buckwheat, amaranth and quinoa

These non-grasses will produce noticeable attractive flowers, which will then wilt. Seed heads will form as the plants die and dry down.

- Flax: July/August. Test by shaking the golden stalks: you should be able to hear the seeds jingle in their dry spherical seedpods.
- Buckwheat: September. Buckwheat seeds don't all ripen at the same time, so wait until about 75 percent of the seeds are hard and then harvest the crop.
- Amaranth: September. "When the birds start pecking at your amaranth plants, it's time to harvest," says Dan Jason of Salt Spring Seeds.

- Quinoa: September/October. Feel the seed heads: if the seeds feel like hard balls between your fingers, they're ready to harvest.

How do I harvest my grain?

Dress appropriately. Remember that many cereal grains have awns — barbed strands that can hook your skin or clothes and work their way in. If you're harvesting grains with awns by hand, wear a long-sleeved shirt. Depending on the height of the grain, you may want to wear pants to protect your legs. If you'll be touching the grain heads, wear gloves.

When harvesting by hand, your first decision will be whether to harvest just the seed heads or the straw as well. This depends on when and how you will thresh your grain. For example, if you plan to store your grain for a while before threshing, seed heads will take up less space then whole stalks. But if you intend to feed your grain into a threshing machine, it may be easier (and safer) to keep the stalks so you have something to

Fig. 15.1: *Small-scale harvesting equipment can be as simple as a pair of scissors and a pillowcase.* (Credit: Sarah Simpson)

Fig. 15.2: *If you want to use your straw for thatch or mulch, cut the stalks at the base, as opposed to just cutting off the seed heads.* (CREDIT: HEATHER MCLEOD)

hold onto. If you want the straw for some purpose (e.g., animal bedding, garden mulch), you can always come back to your garden later and harvest it with a weed whacker or scythe.

For most grains, you can harvest on a small scale with scissors or pruning shears: simply cut the seed head off, or cut the stalk at the length you want. Scythes and sickles work well too, although it's a skill to use either, and it can be difficult to find a teacher these days.

An efficient strategy is to harvest seed heads into a pillowcase or paper bag. Both are breathable, clean containers that can store the seed heads and allow them to continue to dry until you have time to thresh.

Clean your containers between harvesting each batch to prevent the different seed varieties from intermingling, and remember to keep each variety from each seed supplier separate, along with its records.

- Amaranth. Shake or rub the seed heads to loosen the seeds into a bucket, then dry the seeds indoors on trays for at least one week. When the seeds are rock hard, they are ready to thresh.
- Quinoa. Cut the seed heads off and put them in a bucket or spread them on a tarp somewhere warm and dry until you thresh them.

What if I don't have time to thresh right after I harvest, or if the seed heads aren't quite dry enough?

Put the seed heads into a breathable bag, such as a burlap sack, paper bag or pillowcase. Make sure the bag is very clean and doesn't contain other seeds or contaminants. Heritage wheat expert Sharon Rempel recommends placing one label with the grain's information inside the bag, and adding a duplicate label on the outside. If the outside label falls off, you have the other label as back up. Hang the sack in a drafty, dry area for a week or until you have time to thresh the grain.

Alternatively, grain heads can be spread out and left to dry on racks in a greenhouse, as long as the greenhouse isn't humid and is closed to seed-eating rodents and birds.

If you want to dry your grains old-school style, you can stook the grain in the field. Information on how to do this is best obtained by asking an experienced old-timer, but in general a stook (or shock) is made of grain stalks that have been tied into sheaves (bundles). Sheaves are balanced against one another in a teepee shape with the seed heads at the top.

How do I separate the grain seeds from the seed heads?

"Threshing" is the process of loosening the grain kernels from the seed heads and stalk.

Always put safety first: wear a face mask and goggles or safety glasses to protect yourself from dust. Threshing should be done outside or in a large work area (e.g., barn, greenhouse).

Small-scale threshing can be done by hand as needed. You can clean a few cups of grain at a time to make a loaf of bread or meal.

One simple, low-tech threshing method is to place the seed heads in a pillowcase, duvet cover or other enclosed sheet (or to place the seed heads on a tarp, although you're more likely to lose grain kernels this way). Then beat the sheet with a rubber hose, shoe heel, plastic bat or other "flail."

Seed guru Dan Jason threshes his grain with a homemade threshing box: a large rectangular box with one-foot-high sides that has thin wooden slats screwed to the inside to create friction. He drops the seed

Fig. 15.3: *Dan Jason explains the ins and outs of his homemade threshing box at an Island Grains workshop in 2009.* (Credit: Sarah Simpson)

heads inside, then steps in and shuffles his feet to loosen the kernels from the plant. Other designs incorporate a fine metal mesh inside the box to increase friction, although the grains need to be larger than the holes in the mesh for this to work properly.

If you're mechanically inclined, do some Internet research on home-made threshers. One YouTube inventor built a homemade thresher by inserting a large drill into a plastic tub full of seed heads, wrapping a chain around the drill bit and closing up the "drum" with the tub lid. When he turns the drill on, the chain flails the grain.

The possibilities are endless.

Regardless of your chosen method, be sure to clean your threshing equipment and the surrounding area between threshings so as not to mix different seed batches, and keep your records or labels with each batch.

Threshing quinoa

Like cereal grains, quinoa seeds should be thoroughly dry in order to remove them from the plant easily.

Put on a glove to protect your hand (a rubber kitchen glove will do) and run the seed heads through your fingers, scraping the seeds off the stalk and into a bucket.

How do I separate the grain kernels from the plant matter?

Separating the grain seeds from the rest of the plant matter (the "chaff") is called winnowing. Traditionally this is done by placing the threshed grain in a bowl and tossing it into the air, where a breeze can blow away the lighter chaff.

It's surprisingly effective to simply place a large empty container on the ground and pour your threshed grain into it from above, with a breeze (or a fan) blowing away the chaff. It takes about three pours to get quite clean grain.

You can also try using a blow drier (use the "cool" setting so you don't heat the seed too much) or an air compressor to blow the chaff out of a container. Be sure to wear eye protection with this method.

Once you have clean grain seed, make sure it's labeled properly.

Fig.15.4: *Separating the kernels from the chaff is messy. Be sure to do it outside.*
(Credit: Sarah Simpson)

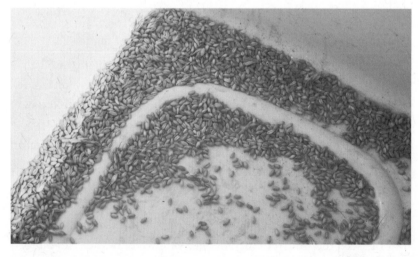

Fig. 15.5: *Wheat that's been winnowed three times using the low-tech bowl-to-bowl method.* (Credit: Heather McLeod)

What do I do with the straw?

The straw and chaff can be added to your compost, tilled back into your soil or used as mulch in your garden. You may want to use your straw for straw bale construction. Wheat or rye straw can be used as thatch, while barley or, preferably, oat straw can be fed to livestock. Wheat straw makes excellent livestock bedding.

What do I do with my grains once I've threshed them?

Put your grain kernels in airtight containers (e.g., milk jug, glass jar) and keep them in a freezer for three to four days. This will kill any insect eggs that may be in your grains.

How do I store whole grains?

Keep a container of whole grains in your fridge for frequent use. The rest of your whole grains are best kept in your freezer, if you have the space.

Never take a container out of the freezer, open it to remove some grain, then put the rest of the grain back into the freezer. Exposing the stored grains to warm air can create moisture in the container.

Oxygen and light will degrade the quality and nutrients in whole grains, so keep your grains in a closed dark, dry container. If you're confident in the dryness of your grains, store the grains in an airtight container, If you're not, store your grains in paper bags or a wood box. To discourage insects from setting up camp in your grain containers, shake or turn the containers at least once every month.

If the grains will be in a garage or shed and accessible to rodents, birds or other animals, put them in a sealable metal garbage can. Rats carry diseases on their feet, hair and saliva, as well as in their urine and feces. Do not eat grains contaminated by rats.

How can I tell if insects have infiltrated my stored grains?

Regularly inspect your stored grains for clumped grains and dust.

If grain kernels appear to have clumped together to form little balls, you may be playing host to a nursery of grain moth eggs. Luckily, the grains can be salvaged. Simply wash the grain in cold water before using it. To dry washed grain so that you can mill it or store it again, spread it on a baking sheet and heat the grains in the oven at the lowest possible temperature, stirring frequently until the grains are completely dry.

Grain dust is a sign of mites. Mite-infested grain should not be eaten by humans or fed to animals. Either clean the container thoroughly or don't use it for food storage again. Mites enjoy humid conditions, so store your grain in a drier, colder place next time around.

How do I get the saponin coating off my quinoa seeds?

Quinoa seeds are naturally coated with saponin, which tastes like dandelion milk. There are machines out there to get the coating off, but small-scale quinoa growers can also use the washing machine method:

1. Run your washing machine through a cycle with vinegar instead of laundry detergent, to clean out any soap residue.
2. Place the quinoa seed in a pillowcase and tie it shut.
3. Using just the water (no soap), run your washing machine through 2 or 3 cycles of cold water with the pillowcase of quinoa inside.
4. Taste the quinoa seed to check that it's saponin-free.
5. When the saponin has been washed off, empty the seed onto trays or cookie sheets and dry thoroughly before storing. If you're concerned that your seeds aren't dry enough for airtight storage, leave them in a paper bag for a few weeks, shaking them daily, to continue the air-drying process.

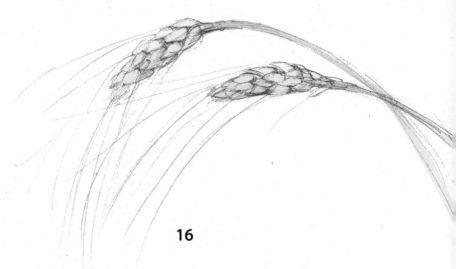

16

Eating Your Bounty: A Feast of Fields

Whole grains contain the majority of basic nutrients essential for life in humans and animals: water, carbohydrates, fats, protein, vitamins, minerals and fiber.

— Joanne Saltzman

YOU'VE PREPPED AND PLANTED, WATERED, WEEDED and harvested. Now comes the fun part: eating the fruits of your labor.

How do I mill grains into flour?

First, consider what kind of flour you want. Most of us are used to white commercial flours, which have had the germ and bran removed: only the starchy endosperm remains. Iron and B vitamins have been added to the commercial flours to replace what was lost in the milling process, creating "enriched" flour.

If you mill your grains at home with a kitchen mill appliance, you will be milling the entire grain to produce a whole-grain flour. You may choose to sift out the bran to create a more mainstream white flour. If eating whole foods is important to you, that will mean eating the entire milled grain.

Milling on a small scale is physically easier these days than our ancestors ever had it. Some home food processors and mixers have grain mill

Fig. 16.1: *It was a proud moment for farmer Brock McLeod the first time he milled wheat he'd grown himself.* (CREDIT: HEATHER MCLEOD)

attachments. North Americans can even purchase home kitchen mills, just as we would a toaster or blender. These specialty appliances are available at a range of prices, from many different suppliers. Some have steel rollers, while others are miniature stone mills.

Alternatively, you can find someone to mill your grains for you. Your local bakery may be able to recommend a nearby mill, if they don't have their own. It just takes a little research.

Finally, you can use other kitchen appliances as makeshift grain grinders. Coffee grinders and blenders can both grind grains, although not usually to the fineness of a proper grain mill.

Part of our Heritage

Under the custom of "thirlage," it was illegal to grind your own grain in England after the Normans invaded in 1066, and in Scotland from the advent of feudalism in the 12th century until 1779. Rebels' home gristmills could be seized and destroyed. This was a handy way to keep the population dependent on its rulers: if there was an uprising, the official mills would close and the people would starve. Milling your own grain not only ensures your personal food security — it's also an act of rebellion.

These days you can mill your own grain, but in North America it's illegal to sift and sell that milled grain as flour (or to sell baked goods made with that flour) unless the milled grain is enriched. Under Canada's *Food and Drug Regulations* and the US's Food and Drug Administration, any white flour sold must be enriched with thiamin, riboflavin, niacin, folic acid and iron. (The addition of vitamin B6, pantothenic acid, magnesium and calcium is optional.) These regulations apply to white flour sold locally and internationally, at any scale.

The move to enrich white flour began in Europe and North America during World War II, when people had limited access to fresh fruits and vegetables. White flour was popular, but it was nutritionally inferior to whole wheat flour or whole grains, both of which still had the bran and germ. Since white flour was consumed by so many people, adding nutrients seemed a logical way to provide the nutrition that people needed at the time.

But keep in mind that milling your grain into flour is only one way to eat whole grains. Grains can be sprouted, cooked like rice or oatmeal, grown into wheat grass, flaked into cereals and much more.

Is flour from locally grown grains different from the commercial flour we get in grocery stores?

For most of us who live outside the major grain-producing states and provinces: yes. These local grains are likely produced in smaller batches by smaller farms than those produced for large commercial sale and export.

Fig. 16.2: *Seeded semolina sourdough made by the folks at Chicken Bridge Bakery in North Carolina.* (Credit: Chicken Bridge Bakery)

Grains that are grown in smaller batches aren't as consistent — not only from farm to farm, but from season to season. Smaller farms may also experiment with non-mainstream grain varieties, such as heirloom wheats, which can adapt to different regions' soil and climates.

Jonathan Stevens of Hungry Ghost Bread dismisses complaints that local flour is not as "consistent" as we're used to using.

"Flour is inconsistent," he says. "Variables exist. We're not scientists; we're craftspeople. We should embrace the challenges that different kinds of flours bring to us. Otherwise, you might as well be on automatic pilot."

Stevens suggests making easier foods, like crackers or flatbread, if you're nervous about the quality or gluten content of your flour.

How do I store milled grains (aka flour)?

First, it's best to mill your grains into flour as you need it because its nutritional value begins to decline after milling. Whole-grain flour will also eventually go rancid, due to the fatty acids in the germ. Shelf life varies, depending on climate, humidity and how it's stored, but whole-grain flour can last as long as nine months.

If you do store whole-grain flour, it's best to keep the bulk of it in an airtight container in the freezer, with some for easy access in the fridge. Just as when storing grain kernels, the flour must be kept dry: don't open a container from the freezer without allowing it to warm up to room temperature first.

How do I replace the usual white wheat flour in recipes with other flour varieties?

New Brunswick's Speerville Flour Mill recommends the following substitutions and taste experiments:

1 cup wheat flour

- 1 cup minus 1 tablespoon spelt flour
- ½ cup wheat flour plus ½ cup buckwheat flour
- 1 cup minus 2 tablespoons rice flour
- up to ¼ cup soy flour plus the rest of the cup wheat flour (reduce the baking temperature by 25 degrees)

Keep in mind that just-milled flour may be stickier than the usual grocery store flour. Wetting your hands with water or cooking oil may make it easier to handle the dough.

What does it mean to "soak" grains?

The first step to cooking with whole grains is to clean them. If you grew the grain or got it directly from the farmer, there may be a lot of chaff to remove. Store-bought grains have been mechanically cleaned and will need less conscientious preparation.

Soaking the grain kernels is an easy way to remove any remaining chaff from your grains. It also neutralizes the phytic acid in the grains, and makes the nutrients in the kernels easier to digest. If you plan to cook your grains, soaking will reduce the cooking time.

Fig. 16.3: *Soaked rye takes less time to cook.* (CREDIT: HEATHER MCLEOD)

Washing grains

First, put two to four cups of whole grains in a large pot or bowl and cover them with at least two inches of water. Stir the grains with your fingers, then pour off any chaff that floats to the surface. Repeat as necessary until the kernels are chaff-free.

Soaking grains

Then soak the grains in lots of fresh water plus one tablespoon of apple cider vinegar or lemon juice. Leave the grains to soak in the fridge for at least 30 minutes, up to 24 hours.

Pour off the soaking liquid and continue with the recipe of your choice. (See the following pages for some suggested recipes.)

What does it mean to "sprout" grains?

Grain seeds are dormant until moisture and warmth activate the growing process. Because of this dormancy, whole grain kernels can last for a very long time if stored properly. One way to make grain seeds edible is to grind them into flour. Another way is to sprout them: to expose the seeds to light, warmth and moisture to trigger the germination process.

Some people with gluten sensitivities can eat sprouted grains and bread made from sprouted grain. (For example, see the recipe for sprouted Essene bread later in this chapter.)

How to sprout grains

Sprout your pre-soaked grains inside a colander set into a bowl, in jars or whatever sprouting method you prefer. If you're new to sprouting and not sure where to start, try this:

1. Put your grains into a large colander that fits inside a bowl.
2. Rinse the grains in the colander, stirring them with your fingers. Then set the colander into the bowl. There should be a gap between the colander's bottom and the bowl.
3. Cover the colander with a loose, dark tea towel to keep out the light while holding in some moisture and allowing air to circulate.
4. Rinse two or three times each day for two to three days, until the kernels form little sprouts the same length as the grain.

Recipes

Grain Berry Salad

Grain berry salad is versatile and delicious, especially when made with rye grains. The whole grains are chewy and filling.

1. In a large pot, cover your pre-soaked grain kernels with water and boil until the grains are soft enough to chew comfortably. Depending on how old the grains are and the variety (i.e., rye vs. wheat), it can take 10 minutes to 2 hours until the kernels are chewy.

2. Drain the cooked kernels into a colander, let them cool a bit, and put them in a large serving bowl.

3. Mince and sauté 1 or 2 onions and a few cloves of garlic in butter until translucent, then add to the salad.

4. Roast/sauté/steam any vegetables you like, and add to the salad. Try diced canned olives, roasted squash, sautéed peppers ...

5. Add small pieces of cheese if you're so inclined (crumbled feta is amazing).

6. Drizzle a salad dressing of your choice over the salad (e.g., creamy garlic, balsamic or honey dill).

7. Add seasonings (salt, pepper, spices, nutritional yeast) and stir.

8. Refrigerate for a few hours or overnight before eating, if possible, to let the flavors blend.

This recipe can be adapted to make a hot casserole simply by cooking the mixture for 30 minutes at 350 degrees. Try adding sautéed mushrooms, chopped cooked kale or chard, grated cheese and/or chopped herbs, then topping with chopped nuts.

Another interesting tweak is to sprout your grains instead of boiling them. Sprouted grains can be used in this salad recipe or to make a cooked casserole.

Sprouted Essene Bread

This dense, chewy "raw" bread is one of the healthiest ways to prepare whole grains and is very tasty, especially when eaten with a soft cheese. Try making this with wheat.

1. Soak, then sprout 2 to 4 cups of grains.

2. If you want to make your bread super healthy, combine in a separate bowl:
   - 1 to 2 cups ground seeds, such as sunflower, sesame and/or pumpkin seeds
   - ½ cup ground flax seeds
   - 1 to 2 teaspoons good quality salt

3. In 2 batches, grind the sprouted grain in a food processor for a few minutes, until the grains are broken up. The "dough" will begin to form a ball.

4. Add the ground grains to the seed mixture and mix well. (Or, if you aren't adding seeds, just combine the salt and grains.)

5. Coat your hands with water or a cooking oil to prevent the dough from sticking to them, and form the dough into 2 or 3 small loaves depending on the size you want.

6. Place the loaves on a baking sheet that's been greased or covered in parchment paper, and bake at 225° F for about 3 hours (or, at your oven's lowest temperature for up to 6 hours).

7. Cool your loaves on a wire rack, then store in an airtight container. Loaves will keep for up to a month in the fridge, and can be frozen for much longer.

Using Flour

Bread (With Poolish)

Poolish is a mixture of flour, water and yeast that's been allowed to sit and ferment at room temperature. If you love the chewiness and nutrition of fermented artisan breads but don't have a sourdough starter to work with, this recipe gives your bread a more complex flavor than just yeast-bread recipes.

1. Make the Poolish:
   - In a large bowl, stir ½ teaspoon yeast into ½ cup room temperature water until the yeast is dissolved.
   - Add ¾ cup flour and stir until the batter is thick.
   - Cover the Poolish with plastic wrap or a lid and let it sit for 2 to 10 hours at room temperature.

2. Mix 1½ teaspoons yeast with 2½ cups warm water and let it dissolve, then stir it into the Poolish.

3. Combine 1 tablespoon salt with 5 cups flour and add it to the Poolish mixture.

4. Mix with a wooden spoon until the dough is too thick to stir.

5. Knead the dough for about 10 minutes.

6. Oil a large bowl and place the dough into it. Cover the bowl and let the dough rise for an hour in a warm place until doubled in size.

7. Cut the dough into 2 or 3 pieces, shape into loaves, cover them and let them rise for another hour, until doubled in size.

8. Bake at 400° F for 45–60 minutes.

Whole Wheat Bread or Buns (With Yeast)

This bread recipe is from the Speerville Flour Mill:

1. Place in a large bowl:
   - 1 tablespoon baking yeast
   - 3 cups warm water
   - ¼ cup honey
   - 3½ to 4 cups of stone-ground whole wheat flour

2. Stir the dough (from the outside inward), folding in air. Cover with a damp cloth and set in a warm place for about 1 hour.

3. Fold into the dough:
   - 2 teaspoons salt
   - ¼ cup oil
   - 3–4 cups white flour

4. Knead for 10 minutes, using 1–2 cups of flour, until the dough is smooth and elastic.

5. Place the dough in a large oiled bowl, cover with a damp cloth, and set it in a warm place for about 1 hour (or until doubled in size).

6. Punch the dough down with your fist until all the air is worked out.

7. Cut the dough into 3 loaves or 24 rolls. Let sit for 5 minutes. Shape the loaves and place them into baking pans or on a floured baking sheet. Cover and let rise for 15 minutes.

8. Cut the top of the loaves with a sharp knife to create steam vents, then brush the loaves with oil, sprinkle them with water, and place them in the oven.

9. Bake at 350° F for 50 minutes, or bake rolls at 375° F for 25 minutes.

Make Your Own Sourdough Starter

It's strange but true: there are natural wild yeasts in the air we breathe. You can "capture" these wild yeasts in your home and make your own sourdough starter to produce truly homemade bread. This kitchen experiment is very rewarding, once you get a bowl of bubbling dough. Be forewarned that, while not labor intensive, the process takes a few days to complete.

1. In a medium-sized glass or plastic bowl, mix together ½ cup flour and ¼ cup lukewarm water.
2. Knead the dough on a flat floured surface for up to 5 minutes.
3. Put the dough back into your bowl, cover the bowl tightly with plastic wrap, and poke 6 small holes into the plastic.
4. Let your bowl sit at room temperature for 12 hours.
5. Add another ½ cup flour and ¼ cup lukewarm water to your starter. Mix well, then place the plastic wrap back on the bowl and let it sit for another 12 hours.
6. Mix in yet another ½ cup flour and ¼ cup lukewarm water, recover the bowl and let it sit for 24 hours this time.
7. If your starter is bubbling after 24 hours: success! You've lured in those wild yeasts and created your very own starter. (If it's not bubbling, toss out the dough and try again.)
8. Now your starter needs to be fed. Mix in yet another ½ cup flour and ¼ cup lukewarm water. Cover it with a new piece of plastic wrap, and this time don't poke any holes into the wrap. Let it sit for 12 hours.
9. Add another ½ cup flour and ¼ cup lukewarm water, cover the bowl and let it sit for up to 8 hours. If your starter is bubbly, it should work in a sourdough bread recipe. If it isn't bubbly, feed it ½ cup flour and ¼ cup lukewarm water every 12 hours for a few more days.

Once your starter is bubbling away, you can keep it in a glass jar in the fridge and feed it once a week, or keep it on your kitchen counter and feed it every 12 hours. It will survive indefinitely, and the sour flavour will become even better with age, as long as you feed it.

Bread With a Sourdough Starter

1. The night before you want to make bread, put 4 ounces of starter into a large bowl. Feed it with 2 cups white flour and 1 cup cold or room temperature water. Mix the dough very well, then put 4 ounces back into an airtight container in your fridge. (This is your starter for future batches.)

2. Leave the rest of the starter-flour-water mixture in the bowl. Cover it with plastic wrap or a loose-fitting lid and let it sit overnight in a cool place.

3. The next morning, add to the mixture:
   - 5 cups flour
   - 2–3 cups cold or room temperature water
   - 1 tablespoon salt

4. Mix the dough well, then knead it on a flat floured surface for roughly 10 minutes.

5. Coat a clean large bowl with oil and place the dough inside. Cover the bowl with a tea towel and let the dough rest for a few hours.

6. At mid-day, cut the dough into three loaves (or rolls), then shape and place the loaves into baking pans or onto a cookie sheet lined with parchment paper.

7. Around 5 p.m., heat the oven to 450° F. Cut the top of the loaves with a sharp knife to create steam vents, then brush the loaves with oil, sprinkle them with water, and place them in the oven. Turn the oven down to 400° F and bake the loaves for about 30 minutes.

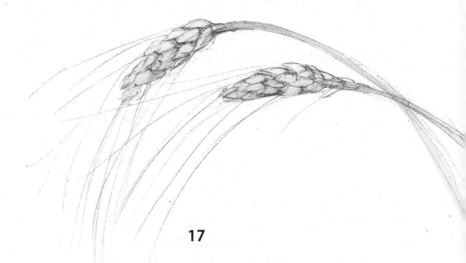

17

Taking the Revolution to the Next Level: Challenge Yourself

You've come a long way, baby. You can grow, harvest and cook your own grain now. But there's no room for complacency. If you're a veteran revolutionary, sunbathing in the afterglow of your own community uprising, we challenge you to take it to the next level.

Here are some ideas to get you started:

Shake Up Your Grain Crops

If you like getting your hands dirty growing grain, there's infinite room for you to play.

- Try planting a variety of wheat or rye both in the fall and in the spring, to see the differences in growth, yield and grain quality. Despite what you've been told about what will and won't grow in your area, you never know for sure until you've tried it.
- Hunt down an unusual kind of grain and see how it does in your own backyard. Less common varieties to grow in North America include Job's Tears, teff (or tef), quinoa, amaranth and einkorn.

Wacky Things to Try When Your Grains Are Growing

- Tough love is the secret to tough healthy plants. Believe it or not, you might actually get a higher yield at harvest time if you subject your growing plants to some stress. In the plant world, "stress" means to make your plants work harder to grow. Try "rolling" cereal grains, or bending them over forcefully, when they're a foot or so high. You don't want to snap the stalks, but rather just flatten them to the point that they need to work to straighten up again.

- Another way to stress your still-green wheat, rye, barley and oats is to mow them, or allow grass-eating animals to graze briefly, once your plants have established themselves well enough to recover. If you plant your grains in the fall, you can safely mow or allow animals to graze once the plants reach a foot or so high in the springtime. Don't mow too short or allow the animals to feast for too long, though. Leave a good six inches of stalk so the plant has enough strength to grow back.

- Grain seedlings aren't just for animals to eat. After all, the wheatgrass shots so popular in health food and smoothie shops are just ... wheat grass, also known as young wheat plants. Head on out to your wheat patch and cut a handful of the green goodness to throw into your juicer. (Nutritionally speaking, it's best to harvest about 14 days after your seeds have germinated.) Quinoa and amaranth seedlings are also edible, nutritious and tasty — but if you harvest more than a few leaves at a time, the plants may not have enough strength to recover. If you sow your quinoa or amaranth seeds too closely together, you can save the seedlings that you thin from the patch and eat the entire plant. Be sure to leave at least one plant for every square foot of growing area.

Experimental Harvesting Methods

Ironically, the methods our forefathers used to harvest grains are now considered unusual. Given that the best way to learn how to properly use these methods is with the close instruction of a practiced expert, many of us will just have to experiment until we figure out the "right" ways of doing things. (The Internet can be very helpful in this area, as well.) If you want to shake things up in the field, consider adding these skills to your repertoire:

- Harvest using a scythe, using a grain cradle if you can find one. While swinging this formidable blade, make sure no one's standing close by. Scythes are significantly more dangerous than golf clubs.
- Use a sickle to harvest. This is a round knife held in one hand that, if not used properly, can disembowel you.
- Once your grain plants have been cut down, collect the stalks into six- to ten-inch bundles and tie them in the middle with a few additional stalks or twine to make sheaves.
- Lean together four sheaves, seed heads up, to make a teepee. Then lean another eight or so sheaves around the teepee to make a "shock." Shocked grains will keep drying down. Check them in about ten days if the weather's hot and dry, or leave them for longer if the weather's uncooperative.
- Once the grain plants are dry enough to store, stack them log-cabin style under cover to make a "rick."

Uprisings in the Kitchen

Once your harvest is safely indoors, there is a host of ways to use those grains to educate others and build connections within your community.

- Track down a "quern," a pair of stones that were traditionally used to manually grind grains in a household. At one time, querns were illegal in Scotland, because they gave villagers an alternative to buying milling services from the village's designated miller: if a villager's illicit quern was discovered, the miller had the right to smash it.
- Learn how to make syrup from barley, rice, sorghum and other grains, and your homegrown pancakes will taste even better.
- Learn how to malt barley ("malting" is just a fancy word for "sprouting") and take a step into the world of home brewing. Although rye and wheat are probably the most obvious grains to be used in alcohol production, most grains can be made into hooch.
- Start a bread club. Band together with friends and take turns making bread each week. If bread making isn't already on your list of skills, ask a pro to teach you and your friends. Maybe they'll even hook you up with

some sourdough starter to give your loaves the benefit of fermentation and help them taste even better.

- Celebrate Bake and Take Month. It's been an annual March event in Kansas and other major grain-producing states since the seventies, and is a great way to celebrate locally grown grains while also connecting with members of your community. Simply choose your favourite grain-based treat (from cookies to quinoa salad) and use it as an excuse to visit with someone you care about. Consider making the recipient of your homemade gift a less-mobile member of your community — perhaps someone in a care facility, or an elderly friend or family member.

Suggested Reading

G ET YOUR GRAIN on with these wonderful books:

Growing Grains on a Small Scale

Small-Scale Grain Raising, 2nd Edition: An Organic Guide to Growing, Processing, and Using Nutritious Whole Grains, for Home Gardeners and Local Farmers, Gene Logsdon, Chelsea Green, 2009.

Homegrown Whole Grains: Grow, Harvest, and Cook Wheat, Barley, Oats, Rice, Corn and More, Sara Pitzer, Storey Publishing, 2009.

The Encyclopedia of Country Living, 10th Edition, Carla Emery, Sasquatch Books, 2008.

Cooking with Whole Grains

Good to the Grain: Baking with Whole-Grain Flours, Kim Boyce, Stewart, Tabori and Chang, 2010.

Ancient Grains for Modern Meals: Mediterranean Whole Grain Recipes for Barley, Farro, Kamut, Polenta, Wheat Berries & More, Maria Speck, Ten Speed Press, 2011.

Super Breakfast Cereals: Whole Grains for Good Health and Great Taste,
Katharina Gustavs, Alive Books, 2002.

The Joy of Cooking, Irma S. Rombauer and Marion Rombauer Becker,
Bobbs-Merrill, rev. ed., 1985.

Amazing Grains: Creating Vegetarian Main Dishes with Whole Grains,
Joanne Saltzman, HJ Kramer, 1993.

Community Ovens

Cooking With Fire in a Public Space, Friends of Dufferin Grove Park, duf-
ferinpark.ca/campfires/pdf/cookingwithfire.PDF

The Bread Builders: Hearth Loaves and Masonry Ovens, Daniel Wing and
Alan Scott, Chelsea Green, 1999.

Revolutionary Books

Coming Home to Eat: The Pleasures and Politics of Local Food, Gary Nabhan,
W.W. Norton, 2009.

The Omnivore's Dilemma, Michael Pollan, Penguin Books, 2007.

Animal, Vegetable, Miracle: A Year of Food Life, Barbara Kingsolver, Steven
L. Hopp and Camille Kingsolver, Harper Perennial, 2008.

The 100-Mile Diet: A Year of Local Eating, Alisa Smith and J.B. MacKinnon,
Random House, 2007. (US edition: *Plenty: One Man, One Woman, and
a Raucous Year of Eating Locally,* Clarkson Potter, 2008.)

Glossary

THESE DEFINITIONS ARE NOT MEANT TO PROVIDE a definitive explanation of these terms but are merely to help you get a basic understanding of grain-related terminology.

Allelopathy: The natural production by a particular plant of biochemicals that affect the growth, survival and reproduction of other plants or insects.

Awns: Stiff, hair-like appendages with tiny barbed ends that grow from the seed heads of some cereal grains. Also referred to as "beards."

Beards: Grains with awns may be referred to as "bearded," while grains without awns are "beardless."

Berries: Grain kernels, which are the "fruits" of grasses.

Bran: The outer layer of the grain kernel. Though high in fiber, the bran is usually removed along with the germ to make commercial white flour.

Bunt balls: When a grain plant is infected with the dwarf bunt fungus, these brown spherical growths replace the grain kernels and may eventually release fishy-smelling spores.

Bushel: A measurement of mass, not weight. The amount of grain that makes up one bushel differs for each grain.

Chaff: Dead plant matter that should be separated from the grain kernels before the kernels are milled or otherwise eaten.

Durum wheat: Wheat with a very high protein content, ideal for pasta making.

Ear: The seed head on the top of a grain stalk. (As in an "ear" of corn.)

Economic clustering An economic model with numerous interconnected industries or sectors.

Endosperm: The seed embryo's food supply while it's germinating and sprouting. The endosperm contains protein and carbohydrates. This part of the grain is ground into commercial white flour, while the bran and germ are removed or destroyed.

Flail: An object that is swung to forcefully separate grain kernels from their seed heads. Traditional flails were made from two pieces of wood separated by a piece of leather or cord. Modern household items such as a running shoe or baseball bat can serve as a flail.

Germ: The embryo of the grain seed, containing protein and oil. It is usually removed along with the bran to make commercial white flour.

Green manure: A crop of plants grown for the purpose of adding nutrients and organic matter to the soil, and/or improving soil structure.

Groats: Grain kernels that have had their inedible hulls removed. If the grains have been mechanically or roughly processed, the kernel may have been damaged and therefore may not sprout or germinate as a seed.

Hull: The inedible layer around a grain kernel. A hull can be paper-thin (for example, around a grain of emmer) or rock-hard and impossible to remove by hand (such as a buckwheat hull). Hulls are also called "husks." Hull-less grains are sometimes referred to as "naked."

Landrace: A grain that can adapt to a particular region's climate, soil and other growing conditions.

Lodging: When grain plants fall down in wind or rain because they are too tall to support themselves. High soil fertility can lead to lodging. Varieties that tend to grow taller are more at risk (e.g., Red Fife wheat, rye).

Malt: Cereal grains that have been sprouted/germinated, then dried.

Naked oats: Oat varieties with a delicate hull that is easy to remove without requiring machinery. Naked oats are sometimes referred to as hull-less oats.

Phytic acid: How phosphorus and energy are stored in the grain kernel. Soaking grains helps neutralize and release the phytic acid, which makes the grains more digestible.

Poolish: A mixture of flour, water and yeast that is allowed to sit and ferment before being used to make bread dough.

Rick: To construct a rick, dried sheaves (bundles of grain plants) are stacked with the seed heads facing inward, and thatch (straw) or a tarp is placed on top to keep off the rain.

Saponin: The invisible coating on quinoa seeds that tastes bitter, like dandelion milk or soap.

Seed: An edible kernel that grows on the top of a grain plant. Grain seeds can be sprouted, cooked, ground into flour or planted.

Seed head: The mass of seeds that grow on the top of a grain plant.

Sheave: A bundle of dry grain plants, like a bouquet, tied with straw or twine.

Shock (aka stook): An arrangement of grain sheaves (bundles) that allows the grain kernels to dry down further after being harvested. Also referred to as a stook.

Smut balls: When a grain plant is infected with smut fungi, these spherical growths replace the grain kernels and may eventually release spores.

Spike: The seed head on a grain plant.

Sprout: A seed that has germinated and has a tap root.

Terroir: A French word often used for grapes or wine that means the distinctive flavor of a crop produced in a specific place.

Thresh: To forcefully remove grain kernels from the rest of the seed head and plant.

Tiller: Can be used as a verb or a noun. Tillers shoot from the below-ground crown of the plant. Each tiller produces its own stalk and seed head. Wheat will send out anywhere from 4–50 tillers, depending on when it's planted, how closely it's planted to other seeds, weather, soil fertility and other conditions.

Triticale: A hybrid variety of grain produced by breeding rye and wheat.

Winnow: To separate grain kernels from the chaff, often by exposing the threshed grains to a breeze or fan so that the air blows away the lighter chaff and separates it from the heavier kernels.

Index

About the Authors

SARAH SIMPSON is an award-winning Canadian journalist. She currently reports for the *Cowichan Valley Citizen*, and her pieces have been syndicated in various major daily newspapers such as the *Globe and Mail* and the *Vancouver Province*. Her coverage of local issues such as the Island Grains project has been recognized in the Environmental Initiative and Community Service categories of the Canadian Community Newspapers Association's Better Newspapers Competition Awards and the British Columbia & Yukon Community Newspapers Association Ma Murray Awards.

HEATHER MCLEOD is the co-owner of Makaria Farm and the co-founder of the successful community grain-growing project Island Grains. She is a passionate believer in re-skilling and founding member of the Renaissance Women, a group dedicated to personal empowerment through re-learning basic skills. Heather has been published in *Small Farm Canada*, *The New Quarterly*, *The Dalhousie Review* and *Room (of One's Own)*, and she writes a regular farming column for *The Winnipeg Review*.

If you have enjoyed *Uprisings* you might also enjoy other

BOOKS TO BUILD A NEW SOCIETY

Our books provide positive solutions for people who want to make a difference. We specialize in:

Sustainable Living • Green Building • Peak Oil
Renewable Energy • Environment & Economy
Natural Building & Appropriate Technology
Progressive Leadership • Resistance and Community
Educational & Parenting Resources

New Society Publishers

ENVIRONMENTAL BENEFITS STATEMENT

New Society Publishers has chosen to produce this book on recycled paper made with **100% post consumer waste,** processed chlorine free, and old growth free.

For every 5,000 books printed, New Society saves the following resources:[1]

24	Trees
2,136	Pounds of Solid Waste
2,351	Gallons of Water
3,066	Kilowatt Hours of Electricity
3,884	Pounds of Greenhouse Gases
17	Pounds of HAPs, VOCs, and AOX Combined
6	Cubic Yards of Landfill Space

[1]Environmental benefits are calculated based on research done by the Environmental Defense Fund and other members of the Paper Task Force who study the environmental impacts of the paper industry.

For a full list of NSP's titles, please call 1-800-567-6772 *or check out our website* at:

www.newsociety.com